T0185598

Mathematics in Mind

The monographs and occasional textbooks published in this series tap directly into the kinds of themes, research findings, and general professional activities of the **Fields Cognitive Science Network**, which brings together mathematicians, philosophers, and cognitive scientists to explore the question of the nature of mathematics and how it is learned from various interdisciplinary angles. Themes and concepts to be explored include connections between mathematical modeling and artificial intelligence research, the historical context of any topic involving the emergence of mathematical thinking, interrelationships between mathematical discovery and cultural processes, and the connection between math cognition and symbolism, annotation, and other semiotic processes. All works are peer-reviewed to meet the highest standards of scientific literature.

More information about this series at http://www.springer.com/series/15543

Kumiko Tanaka-Ishii

Statistical Universals
of Language

Mathematical Chance vs. Human Choice

 Springer

Kumiko Tanaka-Ishii
Research Center for Advanced
Science and Technology (RCAST)
The University of Tokyo
Tokyo, Japan

ISSN 2522-5405   ISSN 2522-5413   (electronic)
Mathematics in Mind
ISBN 978-3-030-59379-7   ISBN 978-3-030-59377-3   (eBook)
https://doi.org/10.1007/978-3-030-59377-3

This Springer imprint is published by the registered company Springer Nature Switzerland AG
The registered company address is: Gewerbestrasse 11, 6330 Cham, Switzerland

Contents

Part I
Language as a Complex System

Chapter 1
Introduction

1.1 Aims

For nearly hundred years, researchers have noticed how language ubiquitously follows certain mathematical properties. These properties differ from linguistic universals that contribute to describing the variation of human languages. Rather, they are *statistical*: they can only be identified by examining a huge number of usages, and none of us is conscious of them when we use language.

Today, abundant data is available in various languages, and it provides a clearer picture of what these properties are. They apply universally across genres, languages, authors, and time periods, in a range of sign-based human activities, even in music and computer programming. Often, these properties are called scaling laws, but the term is not applicable to all of them. Because they are both statistical and universal, we call them *statistical universals*. This book's aims are to provide readers with a review of recent findings on these statistical universals and to present a reconsideration of the nature of language accordingly.

A key representative of previous literature on statistical universals is Zipf (1949). In that study, George K. Zipf described certain statistical universals and considered them evidence of the efficiency underlying language. Prior to that book were other important works such as Yule (1944). After Zipf's book, Herdan (1964, 1956) and Thom (1974) also showed the mathematical nature underlying language. Baayen (2001) presented an important overall analysis of rare words in relation to Zipf's law. Recently, Kretzschmar Jr. (2015) considered the law as evidence of the emergent nature of language and argued its relation to complex systems within the field of linguistics.

Numerous researchers on specific themes related to statistical universals, in the fields of statistical mechanics and computational linguistics, have discovered

---

The original version of this chapter was revised. The correction to this chapter is available at https://doi.org/10.1007/978-3-030-59377-3_25

© The Author(s) 2021, corrected publication 2023
K. Tanaka-Ishii, *Statistical Universals of Language*, Mathematics in Mind,
https://doi.org/10.1007/978-3-030-59377-3_1

other statistical qualities of language that go beyond Zipf's law. Nevertheless, the reports to date are relatively short individual papers and focus on specific topics. A book chapter by Altmann and Gerlach (2016) presented an overview of statistical universals, but it was too brief to cover the complete, scattered nature of the studies. Therefore, the relations among the various findings have not yet been clarified, and the frontier of research into statistical universals remains obscure.

Against that background, this book provides an up-to-date argument on the statistical universals in a larger volume than a research article. Specifically, the aim is to provide researchers in computational linguistics with a consistent understanding of statistical universals. The argument is based on analyzing the mathematical behavior of large numbers of samples, as is often done in fields such as statistical mechanics and complex systems theory. The reason for studying large amounts of data is that it can reveal properties that are invisible when we only study smaller samples. For example, a few tosses of a die result in a short sequence of numbers, but a billion tosses show a new picture of the die's nature. In the long run, an ideal die should give almost the same number of results for each of the six faces. A real die, however, is not a perfectly cubic shape, and therefore, the distribution of tosses will eventually show this bias.

Hence, this book considers how the statistical universals stipulate the characteristics of language. At the same time, it highlights how language deviates from standard statistical behaviors. Indeed, certain statistical universals confirm the expected mechanics, as if language behaves like an ideal die. In these cases, the statistical universals are trivial, yet an important question remains: how does this mechanics stipulate the nature of language? In contrast, other universals deviate from the expected mechanics, reflecting how language also behaves like a biased die. In those cases, the bias might represent some human factor requiring further study to reveal its origin.

The poet Stéphane Mallarmé once compared the action of composition to a throw of dice (Mallarmé, 1897).[1] He pointed out that our use of language can never be free from chance, and his poetic composition highlighted the challenge of this fact. A linguistic act could perhaps be a mixture of chance and choice (Herdan, 1956). Statistical mechanics should partly reveal the nature of the first factor, of chance, as linguistic acts proceed while partly sampling past phrases. The second factor, of choice, is then what drives us to speak: it derives from human intention, as opposed to chance. If that is the case, then identifying the statistical component of the entire phenomenon would reveal the nature of intention.

1.2   Structure of This Book

Figure 1.1 situates this book at the intersection of four subjects. The left part of the figure shows various factors related to language, in particular those underlying the

---

[1] Mallarmé wrote a composition entitled *Un coup de dés jamais n'abolira le hasard* (*A Throw of the Dice Will Never Abolish Chance*) (Mallarmé, 1897).

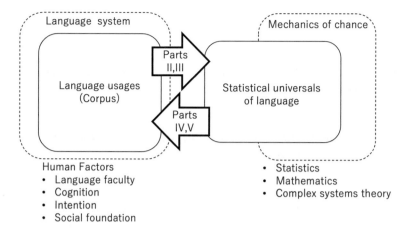

Fig. 1.1 Illustration of the book's structure

usage of a *language system*. These factors include the language faculty, cognition of language, and intention, and the social foundation necessary for language to work. This book mainly considers language usages accumulated in the form of a *corpus*, a large quantity of language data, which Chap. 3 defines in detail. The right part of the figure shows the *statistical universals of language*, which are revealed by certain computational procedures, and the mechanics of chance underlying them. The mechanics of chance is an inevitable consequence of a large number of events involving chance. The book is thus an overview of the statistical universals underlying language, especially in regard to human factors, and how those universals stipulate language.

After Part I, which positions this book within a multidisciplinary academic context, the chapters establish relations between the left and right parts of Fig. 1.1. The first half, in Parts II and III, explains different statistical universals obtained with large corpora, as represented by the rightward arrow in the figure. The first half also discusses the different characteristics of statistical universals for language sequences artificially created by chance.

As represented by the leftward arrow in the figure, the second half of the book considers how the statistical universals explain the nature of language. It focuses on the reasons for the statistical universals, namely how they stipulate the nature of language, or what mathematical and human factors underlie these phenomena. In particular, it examines whether random processes can fulfill the statistical universals of language. A random process approximates language as a sequence of chance events. The results show that certain state-of-the-art processes have the potential to reproduce the statistically universal nature of language, but their capability is currently still limited. This state of affairs implies future directions for understanding language from a mathematical perspective.

1.3   Position of This Book

This book deals with statistical properties of language, as revealed by computational studies on large-scale data. It draws upon fields related to language and computing, and also statistical analysis.

1.3.1   Statistical Universals as Computational Properties of Natural Language

As Hey et al. (2009) assert, data science has become the fourth paradigm of science, and it holds the key to better engineering of large-scale data. Language data is one of the largest and most important forms of big data. Such big data is now being processed to support human linguistic activities. The techniques and methods of language engineering are studied in the field of *natural language processing*. The primary target of the field is thus engineering to provide people with computational assistance for processing language.

Issues in engineering often attract scientific interest. The scientific view of natural language processing is highlighted by the term *computational linguistics*.[2] Moreover, the intersection of computation and language includes other fields that study language by means of computers, including   *quantitative linguistics* and *corpus linguistics*.

In computing with language, we must understand the properties of language from a computational perspective. This book attempts to provide one such perspective. I believe that it not only fulfills a scientific aim but also contributes to the goals of language engineering. One possible engineering objective would be to build computational language models that exhibit those properties. That is, good language models should reproduce the properties of natural language, to better assist language processing.

Over the years, Zipf's law and other related laws have been incorporated in language models. Part II discusses the frontier of studies addressing those traditional laws. The main focus of this book, however, lies rather in the properties described in Part III. Specifically, a property of language called long memory has been quantified more recently by borrowing concepts from complex systems theory. Whether computers can reproduce long memory is an open question in machine learning. Therefore, this characteristic of language must be computationally quantified to clarify the frontier of more advanced language computation. Reproducing only Zipf's law with a language model is not especially difficult, but reproducing all the properties, including those of Part III, remains challenging. Part V describes these issues.

---

[2]In this book, the term computational linguistics stands for both natural language processing and computational linguistics, following convention.

The properties considered in this book hold universally across languages. The fields of study dedicated to language are broader than those using computational means, and the question of universal properties that hold across a variety of languages, or even all languages, has been an important one in the long history of linguistics. Therefore, the statistical universals must be considered in relation to linguistic universals. Thus, Chap. 2 positions the statistical universals within the history of linguistic universals.

Gaining an understanding of universals involves other factors besides the quality of data, because such data is generated by humans. Therefore, this book also takes a cognitive approach in places, by showing how the statistical properties of language relate to linguistic universals and the findings of recent cognitive studies. In this sense, the book partly involves cognitive linguistics, too. Various researchers have chosen approaches based on their own interests and backgrounds. Nevertheless, given the common target of language, the essential questions should be common, irrespective of the disciplines in which they arise. Hence, this book provides one perspective on language, gained through my learning from previous studies bridging those divisions.

1.3.2   A Holistic Approach to Language via Complex Systems Theory

The contraposition of linguistic and statistical universals mentioned in the previous section can be examined in terms of approaches to language from different scales. Figure 1.2 shows the range of language units at different sizes, with a corpus at the top and a sound at the bottom. Linguistics is not always constructionist or reductionist, but a typical book about language proceeds from a microscopic to a macroscopic perspective: from phonemes to words and then phrases. The upward arrow represents this approach. Accordingly, studies in computational linguistics have proceeded from a small unit of morphological analysis to a larger unit of text structure.

Fig. 1.2 Holistic and constructive, the two contrasting approaches to language

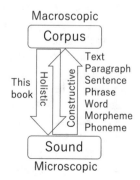

On the other hand, this book takes a holistic approach by examining language through the holistic properties of corpora. The father of modern linguistics, de Saussure (1911), suggested this approach, as follows:

> We should not start from words, or terms to deduce the system. This would assume that terms have absolute values, and the system is acquired only by constructing the terms one with the others. Conversely, we should start from the < system > that works altogether; this last decomposes into certain terms, although this is not at all so easy as it seems.

Researchers following Saussure's line of inquiry have referred to a holistic property of a language system as a *structure*. Although many have sought to determine what that structure is, their findings have been limited to analogies and metaphors. Such analysis is not rigorous enough to be meaningful in processing a large quantity of language data.

Then, what methodology would be appropriate for analyzing such a holistic structure? Language is primarily used by different speakers through individual linguistic acts, the accumulation of which inevitably leads to statistical characteristics. As nobody has ever uttered or written a word by attempting to produce the macroscopic properties of language, such linguistic acts can be better understood in relation to the statistical behavior of large numbers, which is a topic that goes beyond language. The field of statistical mechanics is dedicated to the study of large-scale phenomena in which vast numbers of elements interact at different scales. The consequences of statistical mechanics are commonly described in terms of *limit theorems*, including power laws. When statistical mechanics is applied to a real, large-scale system, the system is called a *complex system*, and the theory developed through study of these systems is called complex systems theory. Thurner et al. (2018) provides an overview of the theory of complex systems.

Complex systems theory has been applied variously to a wide range of natural and social systems. It has seen relatively little application, however, in language. The theories of physics primarily apply to natural systems, and their outcomes should not depend on human interpretation. In contrast, language is characterized as a system of interpretation. Because of this, the main approach to studying language has been to analyze words and sentences in light of some human interpretation of syntactic and semantic roles. Such analysis based on interpretation does not conform easily with the statistical mechanics approach, so studies that treat language as a complex system have remained in the minority. Nevertheless, some researchers in the field of statistical mechanics do study language, and this book owes a lot to work published outside the academic fields dedicated to language studies. To explain this stance, Chap. 3 shows how language can be studied as a complex system.

Statistical analyses of language data have revealed certain universal macroscopic properties. These properties have been attributed as a "mysterious" quality of language, but the actual causality is probably reversed. As this book will argue, it is probable that this mysterious quality is some set of mathematical facts, and that the dynamics giving rise to the universal properties are the precursor of language. Language can be partly characterized by the properties of large numbers. It is thus likely that these dynamics influence the inherent components of language, namely words and grammatical structures. Furthermore, it would be fruitful to know how language can be characterized in comparison with other systems sharing the same precursor.

To highlight the possibility of developing an approach from a statistical and macroscopic view, this book starts from the corpus level and considers the relations between a corpus' properties and those of its elements. The book starts by presuming words as linguistic elements, but later, it shows how words arise partly from global properties. It seems reasonable to say that a corpus influences or even stipulates its elements. In other words, there should be a reflexive dependence between the linguistic elements and the corpus. The organization of the book is hence reversed: it starts from the corpus level and proceeds down to words and phrases.

1.4 Prospectus

The goals of this book are thus to summarize the current understanding of statistical universals and to consider how they might function as a precursor to language, stipulating both its elements and individual linguistic acts. In other words, this book is about the structural, holistic properties of language systems, as found empirically in data. The content is interdisciplinary: it treats computational linguistics from a perspective of language as a complex system.

The book is based on the great insights of various forerunners, with additional findings from my previous studies. Although it is limited by the current state of our knowledge about language, and by my capability of communicating with different audiences, I have tried to cross borders between disciplines.

The prospective audience includes the following readers. For those who study language with computers, the book provides an overview of the global properties of language and how they relate to important notions gained through computing. For linguists, it provides a macroscopic perspective that differs from the perspective of traditional linguistics. For physicists who are interested in language, it provides basic examples showing how the methods of physics can be applied to language and how language is yet another complex system. Finally, for general readers who are interested in language, the book explains the new, emerging frontier of using big data to study and understand language.

For those who are at ease with mathematical formulas, I formally define properties when necessary. Some content involves rigorous formulations, for which the theoretical mathematical background, including proof summaries, is given in Chap. 21. To make the book self-contained, summaries are provided for most of the theoretical rationales. Theorizing through mathematical contemplation often requires making assumptions about the object of interest. As Part III demonstrates, however, language is likely not conducive to simple assumptions. Hence, the book does not presume that arbitrary properties underlie language.

Questions about language tend to attract researchers and students in the humanities. Although this book must invoke mathematical concepts, I have made all possible effort to appeal to a broad audience. I have thus kept mathematical formulas and details to a minimum, although Parts II and III do require an understanding of certain *procedures* used to derive universals. To communicate abstract mathematical concepts that could be difficult for some to digest, I also include simple examples in the main text and in footnotes. Empirical figures and examples are likewise presented to intuitively communicate the meanings of the various properties. I invite those in the humanities to embrace the global message rather than give up because of impenetrable mathematical details.

To make its presentation rigorous, this book focuses on computational aspects. Today, the availability of computational resources has given us greater freedom to describe phenomena even without an underlying mathematical theory. Much of the book relies on this aspect of computation. In other words, the presented statistical universals of language are rigorous in the sense of being computable, with some aspects being mathematical.

As many readers grasp ideas better through examples, the book also reveals empirically discernible properties through a number of large-scale comparisons of corpora. Chapter 22 explains the details of the corpora used in multiple chapters. Skimming through certain figures could give the impression that the illustrated property is only applicable to that example, but the statistical properties introduced here apply to the extent explained in the corresponding sections of the chapters, or to the extent explained in the cited references when the evidence is not directly presented here.

Finally, I should point out that Chap. 20 concisely summarizes the concepts, terms, and symbolic notations used consistently throughout the book. Although these concepts are defined when they first appear in the book, readers can refer to Chap. 20 if they become lost.

Chapter 2
Universals

As this book is about the universal properties of language, this chapter explains and organizes different approaches taken with respect to the notion of *universals*. A universal of language is defined as a property that holds across all kinds of natural language on Earth. The chapters in Parts II, III, and IV consider such properties.

The achievements of linguistics are representative of studies on these properties. The main focus of this book, however, is universals found outside linguistics, in statistical mechanics and related fields. In addition, other approaches could be considered to follow the same train of thought as for the universals of language. Hence, this chapter compiles and reconsiders these approaches.

2.1 Language Universals

Across the history of linguistics, there has been a quest for universal properties that hold across languages, as overviewed in Comrie (1981) and Christiansen et al. (2009). Comrie (1981) categorized approaches to studying universals as either empiricist or rationalist; among the representatives of the latter approach is the work of Noam Chomsky.

With his theory of *universal grammar* (Chomsky, 1995), Chomsky formulated a universal model of human grammar by elaborating the idea of phrase structure grammar (Chomsky, 1957).[1] He considered the human linguistic faculty to be largely inborn, and thus, he proposed rationalist models. Because the phrase structure grammar formulation is mathematical, it has influenced not only possible

---

[1]Phrase structure grammar is often attributed mainly to Chomsky, but the idea originated in the preceding work of the great linguists Leonard Bloomfield and Zellig S. Harris (Blevins and Sag, 2013). Harris's work also appears in this book in other chapters, such as Chaps. 11 and 12, for other major contributions.

© The Author(s) 2021
K. Tanaka-Ishii, *Statistical Universals of Language*, Mathematics in Mind,
https://doi.org/10.1007/978-3-030-59377-3_2

theories of language but also other fields, such as theories of computer program compilers (Aho et al., 1986).

With respect to natural language, however, Chomsky's theories have been controversial. For example, in studies related to childhood language, as represented by Tomasello (2003, 1999), many counterexamples to Chomskian theories have been indicated. Moreover, studies of sentence structure have shown how Chomsky's theory of grammar is far too wide in its description, considering all possible combinations to be parts of language. The instances that appear in texts are rather limited, which raises questions on the quality of the theory's description.

Therefore, in linguistics the widely accepted approaches to studying language universals have been roughly empiricist. As language is both syntactic and semantic, there are corresponding empiricist approaches of each kind. From the semantic viewpoint, Morris Swadesh attempted to list the common words that exist universally in any language (Swadesh, 1971). For example, basic terms such as *I* and *hand* appear in many languages. Swadesh sought to develop a universal set of words that are common to all languages, resulting in lists such as the Swadesh list (2021). Unfortunately, the relevance of his approach has been criticized, because it is difficult to judge whether a word in one language corresponds with another word in a different language. For example, whether the terms for *hand* in English and Japanese really are the same is a difficult question to answer.[2] The question of what is the meaning of meaning is difficult to answer, and so is the related question of whether the meaning of one term is the same as the meaning of another. Therefore, it would be challenging to develop a suitable approach to examine universals from a semantic viewpoint.

In contrast, studies of syntactic universals were originated by American structural linguists and have successfully continued until today. Among other universals introduced in Comrie (1981), two representative examples showed the important properties of language underlying words and syntax. For words, Harris (1955) showed a mechanism that possibly bridges between phonemes and morphemes, which Chap. 11 will introduce as *Harris' hypothesis of articulation*. This book starts by assuming the unit of words, but this is based on Harris's hypothesis, that words partly derive from a corpus. In other words, there is a mutual dependence between words and a corpus: the words constitute the corpus, but the words derive from the corpus. Another of Harris' theories, distributional semantics, is also considered in its relation with statistical universals, in Chap. 12.

In another syntactic approach, Greenberg (1963) indicated a correlation tendency underlying word order, which Chap. 14 will introduce as Greenberg's universal of word order in relation with a statistical universal. In particular, the basic word order of the subject, object, and main verb correlates strongly with the modifier-modified order. Such studies have flourished into linguistic projects to describe the features of languages around the globe.

---

[2]For instance, a cat's *paw* in English is often referred to as a *hand* in Japanese.

The degree to which language follows these properties is an important question, as it indicates whether to accept a property as a universal. There are some seemingly almost trivial universals, such as whether there are vowels in every language, but apart from those, nontrivial language universals do exhibit counterexamples. To more precisely indicate that a universal only holds when taking a statistical perspective, these language universals are called *statistical* (Christiansen et al., 2009).

The counterexamples at the levels of words and phrases deviate from normative usages for various reasons, including convention, mistakes, cases of language transfer, or voluntary artistic choices. This range of counterexamples in the study of linguistic universals could contribute greatly to understanding the possible variation in natural languages around the globe. Furthermore, the universal nature of these counterexamples would be interesting to investigate, because they delimit the potential range of language.

Recently, new approaches have reconsidered the question of language universals (van der Hulst, 2008). Studies have taken a more abstract approach from a more communication-oriented viewpoint. In semantics, the universality underlying vector representations of words across languages has been studied (Lu et al., 2015), and this book considers one such topic in Chap. 12. von Fintel and Matthewson (2008) suggested that Gricean principles (Grice, 1989) are universal. Another study debated whether determiners exhibit universality (Steinert-Threlkeld and Szymanik, 2019). These new approaches have great potential to provide a better understanding of language.

2.2 Layers of Universals

The universals considered in this book are the properties that hold for statistics acquired from large-scale language data. The two opposite approaches to language universals—that is, the microscopic and macroscopic approaches—show that there are different layers of granularity with respect to linguistic units. In particular, clarifying what lies between the microscopic and macroscopic approaches would help situate the statistical universals. Figure 2.1 shows the different layers, ranging from microscopic to macroscopic approaches, and including representative references mentioned thus far. In coordination with Fig. 1.2, the vertical range represents the size of the unit, with the macroscopic view at the top and the microscopic view at the bottom. The horizontal range represents the contrast between empiricist (left) and rationalist (right) approaches. Near the bottom are the Greenberg and Harris universals. They are shown on the left side, because the approach is empiricist. Roughly speaking, the primary interests of linguistics lie in these basic linguistic phenomena including the behaviors of words and phrases.

By increasing the size of the target unit of language, studies based on a similar aim of considering universals have evolved beyond linguistics. At the level of discourse, Foucault (1969) analyzed large archives across different fields and sought

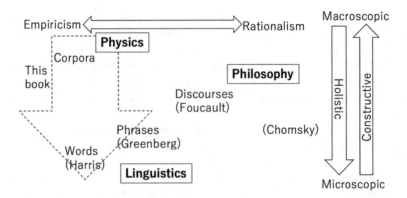

Fig. 2.1 Different approaches to universals of language. The horizontal dimension contraposes the empiricist and rationalist approaches, whereas the vertical dimension represents different scope sizes from macroscopic (top) to microscopic (bottom)

a principle for how concepts evolve in a field. Here, the term *field* indicates an academic field, and thus, a set of discourses involving some academic question. Among many philosophical studies that take a more rationalist approach, Foucault took an empirical approach by scanning through archives of different fields. Therefore, Foucault's approach is situated in the middle of Fig. 2.1.

Interestingly, Foucault emphasized the rarity and variation among utterances playing roles in the formation of a train of thought. Likewise, these aspects of rarity and variation are the essence of statistical universals, and they are considered in Parts II and III, respectively. The statistical universals highlight the rarity underlying the population, and the variation in usages, which appears as fluctuation. What Foucault noticed through rarity and variation not only might derive from human rationality but also could be more deeply rooted in the nature of language. The statistical universals can be considered in parallel with Foucault's thought, but without his notions of history and the humanities.

The statistical laws of natural language consider a target unit larger than Foucault's, situated at the layer of a *corpus* (located at the top of Fig. 2.1). The difference between a *corpus* and Foucault's *field* is that a corpus is an archive of texts that can range across fields. A corpus is not, however, just any collection of text but usually an archive carefully assembled according to policies specified by its creators. One typical corpus consists of newspaper articles and thus includes different topics such as politics, economics, and so on.

In linguistic studies and Foucault's quests, a language sample is analyzed via meaning, so the analysis is related to a system of meaning. Universals are both founded on meaningful units and acquired for a meaningful target. At the level of a corpus, however, words are analyzed devoid of meaning, as will be discussed in the following chapter. The statistical universals of language are investigated with large-scale archives of newspapers and texts that cover multiple fields. Such studies only became possible with big data and powerful computers, and therefore, the statistical universals constitute a new, state-of-the-art understanding that we have been able to gain only recently. As computers cannot understand meaning, words are no longer words in the linguistic sense but treated as mere elements. Each element is presumed to represent some content, but the analysis does not consider that content. Every element is comparable to a molecule, and a text is comparable to a gaseous body. The methodology, therefore, is naturally that of physics, specifically statistical mechanics, or more precisely, complex systems theory. Note that *laws* are primarily an important concept in the field of physics, such as the laws of thermodynamics. Statistical universals of language are closer to that kind of physical law than to Greenberg's language universals, with respect to their different positions toward meaning. Thus, statistical universals of language are discussed more in the field of physics than in linguistics.

Nevertheless, texts are written primarily to produce meaning. At the level of linguistics, universal findings would show the possible range of words and phrases. At the level of discourse, the resulting universals would show the possible range of human ideas and its construction. At the level of the molecular view of texts, the analysis presumes that a target text has some meaning, yet the analysis itself is conducted beyond any meaning conveyed by the text. What, then, would the statistical universals underlying a corpus imply?

The statistical laws considered here present the structure of language that appears universally through statistical analysis. They are non-psychological properties that appear irrespective of meaning. Even then, we have the tendency to consider them as showing some property of a human system, because language is exclusive to humankind. Similar laws, however, have been reported for many other natural systems, as will be discussed later in the book. This fact suggests that the linguistic system could possibly be one such natural system. As explained later, the reason why such properties hold has still not been clarified. Nevertheless, the book takes a stance of considering them as a statistical consequence, a kind of limit theorem, that functions as a precursor of language, as detailed in Chaps. 4 and 15. In other words, such properties have been considered to derive from human linguistic behavior, but there is another view of considering them the statistical premise on which the system of language stands. If statistical universals are such a precursor of language, then what do such properties entail? How do they influence language structures, such as words and grammar? How can language be characterized among other systems that share the same precursor? This book considers these questions.

2.3   Universal, Stylized Hypothesis, and Law

Multiple terms have already appeared in this chapter to indicate properties of language: *universal*, *stylized fact* or *hypothesis*, and *law*. We must first appraise the uses of these terms before redefining the term *statistical universals*.

The field of linguistics has frequently adopted the term *universal* to indicate a property that holds across languages. For example, Chomsky used the term in referring to his theory as universal. The term *universal* also appears frequently in physics as an adjective, modifying the term *law*. The term has been used in both empiricist and rationalist contemplation.

The term *stylized fact* or *stylized hypothesis* is often used in science, including the social sciences, to indicate a property that is known to commonly hold for a certain phenomenon. This term conveys a notion that the reason why the property holds is unknown. In contrast, the term *law* is used in the field of natural science to indicate a property that holds for a certain phenomenon, often for the case when a theory explains why the law holds. As mentioned in the introduction, the properties described in this book are often called *scaling laws*, but the term is not applicable to all of them.

In this book, the properties of language are identified by processing given data as explained in each chapter. These properties are all *empirical* and *inductive*, with no guarantee that they will hold for future texts, although the reality of the statistical universals is that many of them do. When a property does not hold for a sample, it is considered to be an exception or a counterexample. Among these exceptions are two kinds. The first is an exception that violates a property, meaning that the property does not hold in a particular instance. The second forms an exception due to the limitation of the analysis method. For example, microbes cannot become visible without a sufficiently high-resolution microscope. Likewise, if an analysis method intended to reveal a property of a phenomenon is not good enough, then the property stays invisible. This kind of limitation could also apply to the analysis of statistical universals.

Given this background, the true nature of the statistical properties discussed in this book is probably closest to the term stylized hypothesis. Because the statistical properties have few exceptions, however, and because "stylized hypothesis" is a two-word phrase and must often be modified by an adjective (i.e., "linguistic" or "statistical," depending on the field in which a property is found), this book calls these properties *statistical universals* as a short term to indicate statistically universal properties. Nevertheless, when a property is described by two variables related by a power function (defined in the next chapter), the book follows mathematical convention and calls it a power law, even though the actual phenomenon cannot be considered to constitute a law. Relatedly, power properties that have conventional names such as Zipf's law and Heaps' law are called accordingly.

The resulting phrase coincides with the term adopted in linguistics, as mentioned in the previous section, and the term *statistical* indicates that a property almost always holds, though perhaps with some counterexamples. Note that the nature of the counterexamples differs between the linguistic and statistical universals. For linguistic universals, as mentioned in the first section of this chapter, individual instances can easily deviate from expected behaviors. On the other hand, few exceptions occur for statistical universals; as nobody is aware of statistical universals while speaking, deviations are only apparent at the scale of an entire set of samples.

Nevertheless, the two kinds of universals concern the same system of language. Putting the properties highlighted in this book and the linguistic universals together, under the term *statistical universals*, could have the advantage of enhancing the understanding of how they are related. Therefore, Part IV considers how the statistical universals could be connected to linguistic universals that only hold statistically.

Chapter 3
Language as a Complex System

The previous chapter was dedicated to considering the background of this book from the viewpoint of *universals*. It involved notions originating in the humanities, especially linguistics and philosophy. Because language is a system to represent meaning, the analysis to acquire those linguistic universals involved meaning. In contrast, statistical universals, as briefly introduced before, derive from statistical analysis methods. The means—statistics—of this book goes beyond counting and involves statistical mechanics. Hence, this chapter sets out the basis for applying such methods to language.

3.1 Sequence and Corpus

As mentioned in Sect. 1.2, this book considers language mainly via usages accumulated in the form of a *text*, or a *collection* of texts.[1] Specifically, it considers corpora consisting of a large quantity of language data, which could be a single long text or a collection of texts, such as newspaper articles.[2] A corpus is processed as

---

The original version of this chapter was revised. The correction to this chapter is available at https://doi.org/10.1007/978-3-030-59377-3_23

[1] In this book, a text of natural language, used as a sample for consideration by statistical analysis, does not include sets of short, unrelated items presented together, such as a telephone book or a dictionary.

[2] Marcus et al. (1993) distinguished a *corpus* and a *collection*, as follows:

> A *corpus* is a carefully structured set of materials gathered together to jointly meet some design principles, as opposed to a *collection*, which may be much more opportunistic in construction.

They acknowledged that a set of Wall Street Journal articles forms a *collection*, rather than a *corpus*, in strict usage. Recent usages of the term corpus, however, are close to the notion of a large-scale *sample*. Hence, this book uses the term *corpus* to represent a sequence X formed of a text, or a collection.

© The Author(s) 2021, corrected publication 2022
K. Tanaka-Ishii, *Statistical Universals of Language*, Mathematics in Mind,
https://doi.org/10.1007/978-3-030-59377-3_3

a sample of data. A statistical analysis of a corpus seeks to gain knowledge in the form of statistical universals. Hence, this section defines a corpus as a sequence of elements that can be treated mathematically.

3.1.1 Definition of Corpus

This book defines a *corpus* as a sequence, denoted as X, of elements from a given set, W:

$$X = X_1, X_2, \ldots, X_i, \ldots, X_m, \tag{3.1}$$

where i, $1 \leq i \leq m$, is the location index of the sequence, and X_i signifies the ith variable, which takes one element $w \in W$. Based on a function that describes how likely it is for a word w to appear at X_i, the probability of w being the ith element of X is denoted as $P(X_i = w)$. The kind of element in the set W is called a *type*, while the occurrence of an element in the sequence is called a *token*. Therefore, in X, the number of tokens is m, whereas the number of types, $|W|$ is the number of elements of the set W.

Considering a sequence X in this way allows us to examine two aspects:

- the population of elements in W, and
- the sequence of elements of W in X.

This book considers the first aspect in Part II and the second in Part III.

Given a text, the way to define W is not obvious, and both the population and sequence can change depending on the definition of W. In this book, typically, W is a set of words, but there are other possibilities, such as a set of characters.

Part II starts by presuming that the elements of W are words. For English and many other Indo-European languages, W can be acquired fairly easily as a set of elements separated by white space. Many languages, however, notably Chinese, do not separate their elements with white space, and therefore, it might seem that there is no obvious way to define W. For those languages, as long as W consists of a set of word-like elements, the findings presented in Parts II and III will still apply. Later in this book, Chap. 11 will show how words can be acquired by using one of the statistical universals. That analysis suggests a possible way that statistical properties could play a role in generating linguistic elements. Furthermore, the relation between how words in W align to form X and how the sequence of X stipulates W might show how language is reflexive.

Defining language as a sequence X entails some debate about how we should regard language. Above all, language should not be something so colorless as a mathematical sequence. Language has meaning. How should we consider this reality? If we consider language as *a sequence*, then it must conform to some mathematical notions of a sequence. A sample is finite. In contrast, many mathematical analyses assume that a sequence can have an infinite length. Thus, we need to ask: is language finite or infinite? Furthermore, in this book, we suppose that sequences are generated probabilistically. This means considering a sequence as the result of a random process. Is language no more than a random product? Before proceeding to define the basis of the analyses, we should discuss the three points raised here.

3.1.2 On Meaning

The question of what is meaning is perhaps the most difficult in human history. Because only humans can process meaning, an analysis of language considering meaning must rely on human interpretations. Considering language devoid of meaning, therefore, frees the analysis from all of these difficulties that involve meaning.

To understand what Z is, one path is to first contemplate what Z is without the notion of what it is. Then, after gaining an understanding without Z, that understanding is reconsidered by incorporating Z. Philosophers who have contemplated the meaning of meaning tried, in fact, to explain phenomena of meaning by putting meaning aside. This book starts by considering language as a mere mathematical sequence. This approach highlights the properties of language independent of meaning. Then, Chap. 12 applies that understanding to contemplate meaning. Hence, through Part III, the book proceeds without the notion of meaning, but from Part IV onward, it applies the understanding gained through Part III to consider the question of what language is, including meaning.

The most important by-product of this approach is that it justifies the use of computers for analysis. If the meaning of a word does not have to be considered during the analysis, then a computer can do it. This also serves to increase the scale of analysis. In reality, processing natural language involves various difficulties because of meaning. For example, a linguistic word such as "bank" can be ambiguous in its meaning: it can mean a river bank or a financial bank, depending on the context. If the word "bank" is treated simply as an element, however, irrespective of this ambiguity, then a computer can process all occurrences of the term "bank" as the same element. If the one element "bank" should be distinguished in two ways, then an analysis would suggest that there are two sets of usages of that element.

Computers are used for similar processing of language in the fields of natural language processing and computational linguistics, as mentioned previously. Those fields involve engineering, and the outcomes of their studies are expected to be useful for people; hence, their analyses involve meaning. The question of what meaning is in those fields depends on the particular engineering question in the problem setting. Such studies are usually conducted by incorporating the question of what meaning is. For example, one ultimate goal is a machine translation system that can transform a sentence in one language into a sentence with the same meaning in another language. The equivalence of such a translation pair is ultimately evaluated by humans.[3] In a sense, computational linguistics looks for methods that process meaning in a way that appears correct for human interpretation. In contrast, the approach taken in this book initially sets aside the notion of meaning and instead studies the statistical properties of language data before returning to the question of meaning.

3.1.3   On Infinity

As defined previously in Sect. 3.1.1, this book starts by considering language as a sequence. Statistics is a subfield of mathematics, and a sequence in mathematics is often considered to have an infinite length. In that case, the length of a sequence X, denoted as m, is said to go to infinity, i.e., $m \rightarrow \infty$. In mathematics, many conclusions can be drawn for an infinite-length sequence.

For some readers, a finite sequence would seem much easier to handle than an infinite sequence. In mathematics, however, it is often easier to analyze infinite sequences, because a property is often proved to hold when a sequence's length tends to infinity. In contrast, it is difficult to rigorously prove that a finite sequence possesses a certain property. Often such a property remains an approximate one. In this sense, a theoretical consideration of a finite target tends to be more difficult than that of an infinite target.

Then, is language finite or infinite? This is a difficult question to answer. The previous chapters indicated that there are linguistic units of various sizes. Among the small ones are sounds and characters, and these are finite sets in their types. For example, the sound /a/ differs slightly among speakers of different languages; hence, a consideration of individual instances might suggest that the sound /a/ is subject to infinite variations. Nevertheless, the type of the sound /a/ can be considered to represent a limited range of sounds, and similarly, any sound type represents a limited range of sounds. Moreover, each language uses a limited number of sound types; therefore, sets of consonants and vowels are considered finite.[4] A set of characters can also be considered finite.

The linguistic unit of a word is acquired partly through combinations from a set of sounds or characters, C. Let the size of C be $|C|$; then, there are $|C|^n$ possibilities for an n-length sequence. This exponential number grows quickly with respect to n, but if n is finite, then the number of words remains finite. The question of whether n is indeed finite must be framed within a larger perspective. The whole collection of

---

[3]There are computational means of evaluation, such as the BLEU score (Papineni et al., 2002), that compare machine-translated text with human-translated text. Therefore, a human must first translate the text, which of course presumes a human process based on meaning. Furthermore, to evaluate the true quality of the machine-translated text, people must still evaluate the translation quality manually.

[4]Although this book assumes a finite number of types of linguistic sounds, Kretzschmar Jr. (2015) advocates that a phonetic system must be analyzed as a complex system. Research along this line was reported in Torre et al. (2017).

utterances made in human history is finite. But if we suppose that all linguistic acts will continue indefinitely into the future, the question remains: is language finite or infinite?

Noam Chomsky's philosophy is based on the belief that language is infinite. He states in Chomsky (1965) that language must be considered not as some kind of *inventory*, but rather, as a *generative process*. Such an *infinite* way of viewing language is not original to Chomsky, who himself attributed it to Humboldt's philosophy of language (von Humboldt, 1836). Chomsky's formulation of generative grammar produces infinite instances from a finite set.

Similarly, this book considers language to be infinite: as a consequence of the analyses presented in Parts II and III, it shows that language presents the quality of being an open, infinite system. Nevertheless, any sample from a text or a corpus is always finite. Hence, in a way, this book attempts to clarify in what ways language is infinite, through statistical analyses of finite samples.

3.1.4 On Randomness

Whether language is random is a difficult question; finding an answer is one of the major themes of this book. An affirmative answer to this question, saying that language is random, does not match typical notions of language: people speak and write with intention. Under some circumstances, a word has to be that particular word. Still, could that word not have been another one? Is intention itself not a random phenomenon? Such consideration suggests that language might be yet another random process. As represented by Mallarmé's throw of dice in Chap. 1, philosophers and artists have contemplated this question.

Although *randomness* ultimately involves a broad variety of notions, this book considers it mainly in terms of probabilistic occurrences. That is, randomness here primarily means selection of an element by chance from a set of possible elements according to some probability. A random sequence is acquired from consecutive probabilistic choices, via a probabilistic generative process. One thread in this book starts by comparing language with the simplest of random sequences and ends by comparing it with more complex random sequences. The characteristics of language, however, are contemplated through qualities that cannot be described by random sequences. As a consequence of the analyses presented in this book, it becomes clear that language is not yet fully describable by any probabilistic generative process that we know of. Even if such a process is discovered in the future, there is no guarantee that it will be what humans do in linguistic acts, but it could show the conditions that a random process must meet to behave similarly to language.

Therefore, to understand language, this book uses mathematical and computational random sequences for comparative purposes. In this sense, even though language might be a random process, the book's notion of a *random sequence* does not include language.

3.2   Power Functions

Hence, this book considers a corpus as a sequence X. The mathematical preparation for the following Part II continues here.

Given X, suppose that a variable y is empirically measured by some procedure in correspondence with another variable x, where x and y denote two different characteristics of X. Then, we can treat y as a function of x. A key function appearing often in this book is described by the following relation:

$$y \propto x^b, \tag{3.2}$$

where b is a functional parameter. The symbol $\propto$ indicates that y and x are related proportionally by another constant factor a, as follows:

$$y = ax^b. \tag{3.3}$$

The use of $\propto$ highlights the constant b, which governs the essential relation between x and y, as will be seen in the following. From a logarithmic perspective, defining $y' \equiv \log y$ and $x' \equiv \log x$ gives the following equation:

$$y' = bx' + a', \tag{3.4}$$

with $a' = \log a$. Therefore, y' is a simple linear function of x'. This viewpoint is far simpler than considering the relation between x and y, as we can easily observe the linear relation between x' and y'. Its slope is b, which governs the trend of the relation between x and y. Note that a' indicates the degree of vertical shift in the double-logarithmic perspective.

Therefore, to better visualize the relation between x and y, a graph can be plotted on *double-logarithmic* axes, i.e., axes representing the values of $\log x$ and $\log y$, rather than the values of x and y directly. Figure 3.1 shows an example for the case

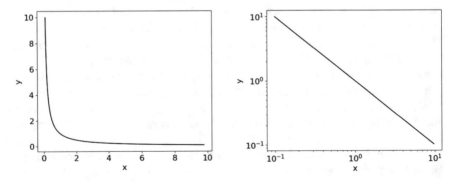

Fig. 3.1 Graphs of $y = x^{-1}$ plotted on normal and double-logarithmic axes

of $b = -1$ and $a = 1$, i.e., the equation $y = x^{-1}$, or, $y = 1/x$, for $x > 0$, which gives $y' = -x'$. The left graph shows the relation between x and y (normal axes), whereas the right graph shows the relation between x' and y' (or, x and y on double-logarithmic axes). In the left graph, the plot is a curve, and the ranges of the axes are limited. On the other hand, in the right graph the line is straight, and the slope b can easily be deciphered by viewing how much y changes for a unit change of x.

For a function plotted on double-logarithmic axes to show a power-law property, the two variables, x and y, must be measurable over wide ranges of values. These ranges should extend across many orders of magnitude (or decades in base 10), because the situation is exponential. To conclude that two variables are actually related by a power law (i.e., a power function), Stumpf and Porter (2012) presented a criterion that the power function should hold for at least *a few decades*. For example, in base 10, the range of each variable should go from 1 through 10, 100, and 1000. Following Stumpf and Porter (2012), this book does not consider power properties that have been reported previously but do not follow this criterion. Also, unless mentioned otherwise, the base of logarithms in this book is 10.

For language, a power function is characterized by its exponent b, which is estimated by fitting data points obtained from a text. Section 21.1 defines two fitting methods, the maximum-likelihood and least-squares methods. There are important issues with fitting, as discussed by Clauset et al. (2009), but those issues are beyond the scope of this book. Accordingly, b is estimated by using the maximum-likelihood method when applicable[5] and by the least-squares method on a logarithmic scale otherwise. Section 4.5 further considers the issue of fitting.

3.3   Scale-Free Property: Statistical Self-Similarity

A power function has a *self-similar* nature. Self-similarity is a property in which the whole includes itself at smaller scales. For example, consider the fractal Koch curve shown in Fig. 3.2. A Koch curve includes itself at a smaller, one-third scale.

Fig. 3.2 A Koch curve, an example of a completely self-similar object

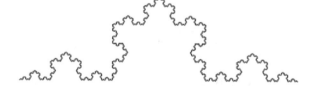

---

[5]The maximum-likelihood method is applied when the variable to be fitted can be naturally defined in the form of a probability function.

The power function described by $y \propto x^b$ includes itself as a part and has a self-similar property. Enlarging x by a factor of k produces the following transformation:

$$y \propto (kx)^b \propto k^b x^b \propto x^b. \tag{3.5}$$

The relation between y and the enlarged x stays invariant, preserving the same power exponent b. Unlike a power function, other functions change their shape in this transformation. For example, in the case of the exponential function $\exp(x)$, enlarging x by a factor of k results in $\exp(kx)$, a function with a different shape. A power function thus has the quality of preserving a fundamental functional form with respect to scale changes. This quality is called a *scale-free, scale-invariant,* or scaling property. An observation for which a power function applies is thus scale-free. The scale-free property is a kind of self-similarity, because scale independence implies that a system has a smaller copy of itself within. The notion of self-similarity is a more general idea of the whole including itself, whereas the scale-free property highlights situations in which there is an invariability with respect to scale. Thurner et al. (2018) provides more formal definitions of *scaling* and *self-similarity*.

When a power function holds universally across samples, the property described by that function is often called a *power law*. This implies that the scale-free property holds universally, and the property is called a *scaling law*. Many of the statistical universals discussed in Parts II and III exhibit power laws, which means that language has self-similar aspects.

A Koch curve is infinitely self-similar, in that the object includes itself at smaller sizes by a factor of $1/3^k$, with $k = 1, 2, 3, \ldots, \infty$. On the other hand, real self-similar phenomena are not infinitely recursive. The degree of nesting—that is, the number of instances of the whole included in itself—is limited partly because a real whole has a finite size and cannot be divided up infinitely. This is one reason why the power functions considered from Part II onward present some bias, such as not being completely linear in a log-log scale. In addition, this bias could derive from other, human factors. Language is such that self-similarity is deemed to hold only roughly when language is analyzed statistically. When statistical analysis methods reveal a rough self-similarity underlying a phenomenon, we will call this property *statistical self-similarity*. Hence, the book aims to clarify how language differs from a perfect fractal and to understand the human factors underlying this difference.

3.4 Complex Systems

This book treats language as a kind of complex system. A rough definition of a complex system is one whose elements interact in a complex manner and show a holistic behavior that cannot be explained by breaking the system down into elements and studying their individual behaviors (Thurner et al., 2018). Many natural and social systems are complex systems. Various notions of complexity, including reflexivity, variation, nonstationarity, long memory, and nondeterminism, involve the properties of language.

A common outcome of a complex behavior is often observed as some global statistical behavior. Such global attributes of a system are often used to define a complex system. Following the viewpoint proposed in Bak et al. (1987, 1988), this book defines the term complex system to mean a system with the following two macroscopic properties, which are often seen among complex systems:

- a scale-free property underlying the population, and
- long memory.

Parts II and III, respectively, discuss these two properties. The previous section defined the scale-free property, and the first definition above suggests that it underlies the population. The second property, long memory, is an important, common property of complex systems, and in the case of language, two portions of a language sequence that are far apart have some dependence on each other. The dependence could derive from any source, whether syntactic, semantic, or pragmatic.

The term long memory is related in turn to three other notions: *long-range correlation*, *fluctuation*, and *long-range dependence*. *Long-range correlation* is a term from physics, indicating that two sequences of events separated by some distance are correlated. Chapter 8 mathematically defines this notion. *Fluctuation* is discussed in Chap. 9 and indicates how elements tend to appear in clusters, which is one cause of long memory. Lastly, *long-range dependence* in this book[6] signifies the grammatical dependence among words that are not next to each other, such as the subject's dependence on the verb. Chapter 14 describes how grammar serves to separate related words at a certain distance.

Bak et al. (1987, 1988) showed how a dynamical system with spatial degrees of freedom naturally evolves into a complex system characterized by three properties (Jensen, 1998; Pruessner, 2012). Besides the first two properties mentioned above, the third property, called the $1/f$ property, underlies the frequency of signals.[7] Because a language sequence is discrete and nonnumerical, it is not trivial to consider how to define its signal frequency. Chapter 7 investigates a related property underlying return intervals. Therefore, we will also examine the third property, but not in the sense of a continuous, numerical time series.

3.5 Two Basic Random Processes

The chapters of Parts II and III are organized to explain particular statistical properties of language. Each chapter provides an example as evidence, taken from the literary text of *Moby Dick; or, the Whale* by Herman Melville, which was

---

[6]In other fields such as machine learning, the term signifies the entirety of long memory.

[7]This refers to the property of a power spectrum having a distribution defined by $1/f$, the inverse of the frequency.

previously used in individual articles to obtain underlying statistical universals. This well-known masterpiece is about a ship captain pursuing a whale. It begins with the following sentences:

Chapter 1. Loomings.

Call me Ishmael. Some years ago—never mind how long precisely—having little or no money in my purse, and nothing particular to interest me on shore, I thought I would sail about a little and see the watery part of the world.

As mentioned above, we can consider this text as a sequence of words, $X = X_1, X_2, \ldots, X_i, \ldots, X_m$, where X_i represents one element of the sequence. Each element derives from a set of words, W, i.e., $X_i = w \in W$. All the types, i.e., different words, in the text appear in W, whose size is denoted by v. For *Moby Dick*, the above sample begins with the words "Loomings", ".", "Call", "me" $\in W$, giving the elements X_1="Loomings", X_2=".", X_3="Call", and X_4="me". The text length, i.e., the number of tokens, is $m = 254,654$, and the number of different words is $v = 20,472$. Alternatively, we can also view the text as a sequence of characters. In that case, X_1="L," X_2="o," X_3="o," X_4="m," etc., with $X_i = c \in C$, where C is the set of characters. Then, $m = 1,765,145$ and $u = 78$, where u denotes the size of the set of characters for *Moby Dick*, including both alphabetic and other symbols.

In the field of computational linguistics, words are often lemmatized and analyzed in their root forms. Verbs conjugate, and nouns take plural forms. For example, "calls" and "called" derive from the same word, "call." Another issue is how to process symbols such as commas and periods. They are often ambiguous: for example, a period may stand for a sentence stop or a decimal separator. Lemmatization and symbol removal often require human interpretation. When machines perform this preprocessing, there is a question of accuracy, as some languages require nontrivial algorithms. In fact, while machines can preprocess some major languages almost 100% correctly, they cannot yet do so for all the languages on Earth. In particular, starting in Chap. 5, statistical universals are verified for more than 1000 texts in 14 languages, but that is only a fraction of the world's languages. Preprocessing tools have not yet been developed for some languages, and the consequent errors all involve the question of meaning.

Above all, as stated at the beginning of this chapter, the analysis in this book does not concern meaning. Statistical universals show how a language system follows statistical consequences that are primordial to language. The techniques of separating symbols and words and lemmatizing word forms already presume language to be a meaningful system. Before delving into the question of what qualifies linguistic elements, the statistical behavior of elements must be verified for a rough set of elements. The set might be noisy, but language is already noisy in various ways, without a clear definition of what its elements really are, and yet the language system possesses statistical universals. In this book's approach, therefore, each of the derived forms of words and symbols is considered to be a type.

One way to understand the signification of statistical universals of language is by comparing real texts with simple random sequences. Part V thoroughly considers such random sequences, but even before that, it is worthwhile to compare *Moby*

Dick with random sequences. To this end, the book compares *Moby Dick* with two random sequences. The first sequence is a *shuffled text*, acquired by shuffling the words of *Moby Dick*. Specifically, a shuffled text is obtained by randomizing the order of words in a text, resulting in a sequence of words whose order in the original text is destroyed completely. Such text can be obtained, for example, by a procedure that repeatedly samples two words in the original sequence randomly and exchanges them.

Every time we shuffle *Moby Dick*, the output sequence will be different. From a vocabulary perspective, W remains exactly the same as for the original text, but the alignment of words in the sequence X is completely destroyed. The following is a sample of shuffled text for *Moby Dick*:

upon of so . more eastern of desired howled and label powerful 's of leaving , these Long

Note that this shuffled text sample is close to an independent and identically distributed (abbreviated as i.i.d.) sequence. In such a sequence, all the words are independently sampled from the same identical information source of *Moby Dick*. The original source derives from the author, Melville, who wrote the text with words that depend on one another. In an i.i.d. sequence, however, this dependence is destroyed, and words are sampled independently. The shuffled text is, in fact, only an approximate i.i.d. sequence, in the sense that there is a global constraint that the vocabularies of *Moby Dick* and its shuffled version must be equal.

The second random sequence is acquired by sampling characters independently from *Moby Dick*, one after another, to form a text. The result is a sequence of characters (or symbols) followed by a space, another sequence of characters followed by a space, and so on.[8] The sample text is acquired by continuing this process to reach a certain number of characters, as seen in the following sample:

M.siltw, d gov e eo o ls ihhaoeceh'tkt o

Note how this random sequence is like the output of a monkey who has been trained to type according to the distribution of character occurrences in *Moby Dick*. If we consider a space to delimit words as in *Moby Dick*, then a monkey text has elements that correspond to words, but they are far from real words in English. From here on, we call this kind of sequence a *monkey text*. There are other variations for such texts based on monkey typing. Chapter 4 considers a cruder monkey text in which all characters except spaces occur uniformly. Another variation is a character-shuffled sequence in which the number of characters equals that of the original. In this case, the character frequency counts are exactly the same as the original, which is not the case for a monkey text. Indeed, in comparison with a shuffled text, a random sampling of a text is less mathematically constrained with respect to the original text. As such, there are various ways to construct random texts, even for

---

[8] When consecutive spaces occur, they are considered to be one space. In that sense, a monkey text is another approximate i.i.d. sequence.

only approximately i.i.d. sequences from *Moby Dick*. Thus, this book primarily uses character-sampled text, called monkey text, as defined above, to represent an i.i.d. sequence of characters.

Note that these shuffled and monkey texts are random texts, aligned only by chance. This inevitably suggests the ultimate question about random texts: could a random text duplicate the original text of *Moby Dick* only by chance? This question has plagued fields ranging from literature to statistical mechanics. The relation of randomness to language, or the question of how randomness involves language, has also intrigued various artists, such as Mallarmé, as mentioned at the end of Sect. 1.1.

The infinite monkey theorem[9] provides an answer to this question by stating that an infinitely long random sequence *will* eventually produce the desired masterpiece, as long as the probability of that sequence is not zero. For a crude monkey text with $u = 78$ characters and length $m = 1,765,145$, however, there are u^m possible random sequences. Therefore, the probability that a monkey produces the exact desired text is incredibly small indeed. How can we statistically characterize the original magnum opus, *Moby Dick*, in comparison with such shuffled and monkey texts? Do these random texts exhibit the statistical universals of language?

---

[9]Note that the infinite monkey theorem is typically considered w.r.t. Shakespeare, rather than *Moby Dick*.

Part II
Property of Population

Chapter 4
Relation Between Rank and Frequency

Part II mainly considers the characteristics of a population of linguistic elements, such as words. A word has a frequency in a text, and the vocabulary of the text forms a population, which this part analyzes.

The population differs, of course, according to the type of elements that we see in a text. This chapter starts the analysis with words, but we will consider other elements, such as characters, later in this part. Words are typically elements separated by white space in a text for languages such as English. On the other hand, many writing systems do not separate words, but for those languages, Chap. 11 will show how such elements can be acquired merely by using the statistical properties underlying a sequence. With such elements, the basic property explained in this chapter applies to any language. This property of text seems to have been noted by Jean-Baptiste Estoup in 1912 (Petruszewycz, 1973), but it was first studied extensively by Zipf (1949).

4.1 Zipf's Law

Every word appears in a text with a frequency, where frequency here means the number of occurrences in that text. These frequencies depend on the type of element and form a distribution for a set of elements. For example, in *Moby Dick*, "whale" appears 783 times, while "ship" appears 451 times. Then, a word has a rank according to its frequency. For example, "whale" is the 38th most frequent word in *Moby Dick*, and "ship" is the 67th. Let a word $w \in W$ have frequency f and rank r according to the frequency. The ranks of "whale" and "ship" are thus 38 and 67, respectively. As both r and f are numbers, we can consider a functional

The original version of this chapter was revised. The correction to this chapter is available at
https://doi.org/10.1007/978-3-030-59377-3_23

K. Tanaka-Ishii, *Statistical Universals of Language*, Mathematics in Mind,
https://doi.org/10.1007/978-3-030-59377-3_4

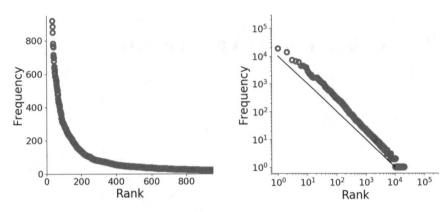

Fig. 4.1 Frequency with respect to the rank in order of frequency for the text of *Moby Dick*. The left graph shows the plot for small r and f on normal axes, whereas the right graph shows the plot for the entire vocabulary on double-logarithmic axes. In both graphs, the red points were measured for the actual text, whereas the black line in the right graph represents a power function with an exponent of -1 (not a fitting function)

relation between them. The resulting analysis therefore concerns the *rank-frequency distribution* of words.

For example, the left graph in Fig. 4.1 shows the relation between f (vertical axis) and r (horizontal axis) for ranks up to a little less than 1000 for *Moby Dick*. The graph shows only the first part of the plot, because the entire plot would be too large to show on normal axes: the vocabulary size of *Moby Dick* is $v = 20,472$, so a complete plot would practically coincide with the x and y axes.

A better picture is produced by taking the logarithms of the rank and frequency, as explained in Sect. 3.2: the right graph is able to show the whole plot for the entire vocabulary. The red points represent all the actual words and exhibit an almost straight alignment. In other words, the frequency f and rank r have a relation close to the following formula, with a and η being constant parameters:

$$f = ar^{-\eta}. \tag{4.1}$$

As explained in Sect. 3.2, this is a power function. The frequency is therefore power distributed with respect to the rank, with some slight deviations from an exact power function. Defining f_{max} to be the frequency for the word of the highest rank, $r = 1$, gives $a = f_{max}$.

The value $-\eta$ is the slope of the plot. In this book, η consistently denotes the negative of the slope of a rank-frequency distribution, as included in the summary of notations given in Chap. 20. For comparison, the black line just beneath the red points has a slope of -1; i.e., it represents the relation $y \propto x^{-1}$, and thus, it shows that the plot corresponds to an exponent close to -1. This black line represents a *harmonic* function, which appears frequently in relation to complex systems theory. It suggests a property that the second-ranked word has half the frequency of the most frequent word, i.e., $f_{max}/2$, and that the kth-ranked word has frequency f_{max}/k.

Zipf (1949) originally indicated this property that the rank-frequency relation of every text forms a power law with a slope of -1, and it is thus called Zipf's law. He thoroughly analyzed various datasets including both language data and data from sources beyond language. Ever since, numerous reports on Zipf's law have appeared in statistical mechanics and various language-related fields. The following list summarizes the consequences of Zipf's law in relation to language:

1. The value of η is approximately 1.
2. Zipf's law is universal. It applies to any text, regardless of the genre, language, time, or place.
3. Zipf's law applies to not only natural language data but also data related to human language, such as music and programming language source code.
4. The law is a rough approximation, and both the heads and tails of distributions often deviate. The plots of some texts show a convex tendency. Furthermore, certain kinds of texts show a large deviation from a power law.
5. Changing the elements from words to morphemes or characters changes the shape of the plot.
6. There are other power laws related to Zipf's law.

This chapter elaborates on point 1. Chapter 5 then discusses points 2–5, and point 6 appears in Chap. 6.

4.2   Scale-Free Property and Hapax Legomena

In what way does Zipf's law characterize text? For any consecutive portion taken from *Moby Dick*, the rank-frequency distribution appears as a power function with $\eta \approx 1$. For example, Fig. 4.2 shows the rank-frequency distribution for the first 1/16th of *Moby Dick* in green, together with the red plot already seen in Fig. 4.1 for

Fig. 4.2 Rank-frequency distributions of the first 1/16th of *Moby Dick* (in green) and the whole text (in red). The black line represents a power function with an exponent of -1

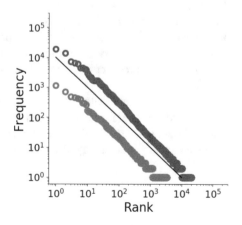

the whole text. The overall plot shifts closer to the origin of the graph but remains parallel to the original plot in red, with a slope of about -1.

The green plot here represents the first 1/16th of the text, but any arbitrary portion taken consecutively from anywhere in the text would produce similar results. This lack of dependence on the size of the text portion is called the *scale-free* property; as explained in Sect. 3.3, it is a kind of self-similarity that is characterized by a power function.

What exactly does this power function mean? Zipf's law indicates that the population distribution—if considered in terms of the frequency among words in rank order—is preserved for any word population of a text. One important characteristic of this population is the number of rare words. Kretzschmar Jr. (2015) called this the "80/20 rule" to represent how 80% of word types yield only 20% of word tokens. The left graph in Fig. 4.1 shows that the vocabulary includes many rare words. Indeed, among all the word types of *Moby Dick* ($v = 20, 472$), almost half occur only once, including examples such as "white-fire" and "weazel". Such words that occur only once in a text are called *hapax legomena*. The proportion of hapax legomena in a text is usually around half, roughly ranging from 40% to 60%. Zipf's law suggests that this proportion remains the same regardless of the text size. Also among the rare words are those that occur twice, three times, and so on, which are fewer in number than hapax legomena but each occur more frequently. Zipf's law implies that this population of rare words in a text can be estimated from that of the frequent words, because the vocabulary population is invariant with respect to the text size.

Nevertheless, rare words appearing in texts are statistically problematic, because their usages cannot be sufficiently observed. For example, each of the hapax legomena occurs only once, making it difficult to estimate their true probabilities. The problem of how to treat the occurrence probability of rare words is an essential one in language modeling in the field of computational linguistics, to which this book returns in Chaps. 10 and 17.

This abundance of rare words shows how a text is an *open* system, in which the vocabulary increases with the text size. Chapter 6 more explicitly examines the degree of increase with respect to the text size. In many basic mathematical phenomena, a set is often considered to be *finite*, as part of a *closed* system. A language system, however, is difficult to put in a finite setting, and this nature of being *infinite* with respect to the vocabulary is the primary property that we should consider.

Overall, Zipf's law states that new words are constantly being added to text, among words of other frequencies, such that the population follows a power law. As will be shown in Chap. 5, this power law is universal, holding for any text authored by anyone, at any time, at any place, and in any language.

4.3   Monkey Text

To what extent this universality should actually be surprising requires careful consideration. One approach is to see whether the property holds for random sequences. The previous chapter defined two such sequences: a shuffled text, and a monkey text. For a shuffled text of *Moby Dick*, the result is exactly the same as that shown in Fig. 4.1, because all the words appearing in *Moby Dick* also appear just as frequently in the shuffled text.

The monkey text raises a more interesting question. It should largely differ from the original text, as it hardly shares any words because of the randomization. As noted in Sect. 3.5, a monkey text has elements, delimited by spaces, that can be considered to correspond to words, and these elements can be used to investigate the rank-frequency distribution. The yellow plot in the left graph of Fig. 4.3 shows the rank-frequency distribution of the monkey text. Note that the figures in this book consistently follow a convention of using yellow for the results of analyses on random texts.

Despite the large difference between the texts, the distribution appears very close to that of the original text. Note that similar analyses appeared in Li (1992) and Bell et al. (1990). This shows that Zipf's law is not just a characteristic of natural language but a property of another, larger set of sequences that includes natural language.

Still, there are differences from the original result for *Moby Dick*. The plot for the monkey text has a crooked region at the beginning (the head), which does not exist in the original *Moby Dick* plot. This crooked feature, however, also appears for

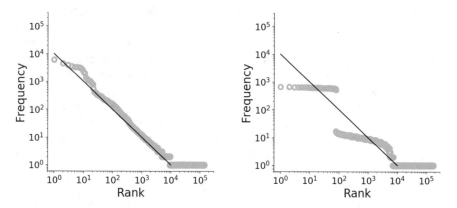

Fig. 4.3 Rank-frequency distributions of two monkey texts. The left graph is for the monkey text defined in Sect. 3.5, which was produced by random sampling of characters from *Moby Dick*. The right graph is for a cruder monkey text in which characters occur with a uniform probability of $(1-a)/(u-1)$, where u is the number of characters and a is the probability of a space. Parameters such as the text length and space probability were set according to the original text of *Moby Dick*. Each black line represents a power function with an exponent of -1

certain real natural language texts. Another difference is that the overall heights of the points are lower for the monkey text, because of the increased number of hapax legomena. Indeed, the tail is longer than that of the original. Such a tendency is not specific to this random sample; it is shared by any sample of monkey text generated by the procedure explained in Sect. 3.5.[1]

To further see what kinds of sequences would roughly obey Zipf's law, we could generate a cruder monkey text. For the monkey text defined in Sect. 3.5, a monkey hits keys according to the frequency distribution of characters in *Moby Dick*. To make a cruder text, the monkey could hit all non-space characters according to a uniform distribution. For example, the monkey hits the space with some constant probability defined by the proportion of spaces in the original text of *Moby Dick* (220,588 spaces out of 1,221,771 characters). It then types the remaining 77 ($= u -$ 1) characters with another probability, which is equal for each of those characters. It thus produces a sequence of characters, a space, and then another sequence of characters, a space, and so on.

The right graph in Fig. 4.3 shows the rank-frequency distribution of this cruder monkey text. The result initially appears rather different from the distribution of the original *Moby Dick*, as the plot appears to be a discontinuous step function. Such steps appear because elements of the same length appear almost the same number of times. Comparison with the black line representing the harmonic function, however, shows how the plot globally follows the power-law tendency.

This last result is actually a theoretical fact. George Miller proved analytically that monkey typing generates a power-law rank-frequency distribution (Miller, 1957). (See Sect. 21.2 for the proof.) The step function is mathematically shown to be smoothed when the probability is not uniform across characters (Conrad and Mitzenmacher, 2004), as seen in the left graph of Fig. 4.3.

These two types of monkey texts demonstrate how the family of monkey-typed sequences nearly obeys Zipf's law. Therefore, Zipf's law does not characterize natural language but is instead a property of a larger set of sequences. Miller's proof shows that the property holds for a set of sequences acquired by randomly aligning elements that are each a member of a finite set. Zipf's law thus holds for a set of sequences that includes this set.

Figure 4.4 illustrates this situation. On the left is a finite set. As discussed in Sect. 3.1.3, a set of sound types is an example of a finite set. The right side of the figure shows a set of sequences for which Zipf's law roughly holds. This set includes sequences of natural language, in addition to various random sequences, including a monkey text and a shuffled text. Part V will consider some other random sequences.

The set of sequences delineated by the dashed oval in Fig. 4.4 is one subset of the above set. This set is acquired from combinations of the finite set shown on the

---

[1]Others, such as Ferrer-i-Cancho and Elvevåg (2010), previously indicated such differences between some of the simplest random texts and the original text. The population of monkey text can also be shown to differ from that of natural language text by different ways from those used in these previous works, as shown later in this chapter, in Chaps. 6 and 13. It is important to know first, however, that monkey texts analytically produce a power law.

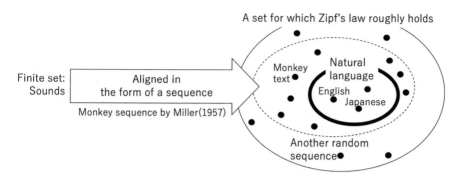

Fig. 4.4 A set of sequences for which Zipf's law holds

left. Alignment of the elements of the finite set generates a set of sequences. If we segment a sequence into portions (for example, with a specific element such as a space) and call each portion a word, then the number of words increases rapidly with respect to the word length. Natural language texts are part of this set, because they are acquired by aligning the sounds forming the finite set. The dashed oval set also includes monkey sequences acquired by aligning elements randomly, as proved by Miller.

Natural language sequences, monkey texts, and many other random sequences share the property roughly presented by Zipf's law. They possess hapax legomena in large proportions, and the shape of the rank-frequency distribution does not depend on the text size. This implies that this property is not specific to language. Rather, language is only part of the set in which Zipf's law roughly holds. If Zipf's law is reproduced by a monkey text, then is it trivial and no longer important? Many have argued against this notion. For example, in the field of linguistics, Kretzschmar Jr. (2015) advocated that an approach to language from the perspective of complex systems is promising because it captures the emergent nature of language. Furthermore, in the field of psychology, Piantadosi (2014) provided an argument that emphasized the importance of Zipf's law and he showed psychological evidence in support of the law. This book also opposes the notion that Zipf's law is trivial.

Even if Zipf's law were mathematically trivial according to Miller's proof, the question of how such a property stipulates language is a different problem from whether the law is reproducible. This is an important question for understanding language. This question suggests the possibility that Zipf's law underlies language as a premise; if so, then it may have something to do with the formation of language. Part IV discusses this possibility.

Furthermore, if language is only part of a larger set, naturally we would want to know the inherent characteristics specific to language sequences. Monkey or shuffled sequences must be easily distinguishable from natural language sequences. Furthermore, even if we have another random sequence for which Zipf's law roughly holds, it should also be easily distinguishable from a natural language sequence. Such distinctions have the possibility to characterize natural language, and they could be applied in the form of a statistical *model*. If such a model does not fulfill Zipf's law, then the corresponding sequence is situated outside the outer set in Fig. 4.4. On the other hand, if the model does fulfill the law, then we can apply similar verification for the other statistical universals considered later in this book. Then, if a model can reproduce all the statistical universals of language, it could be a good model for language engineering. This book takes such a view: after covering all the statistical universals, it evaluates various natural language models in this way in Part V.

The rest of Part II focuses on Zipf's law, and on verifying the characteristics of natural language in more detail from the perspective of Zipf's law.

4.4 Power Law of *n*-grams

In fact, the difference between *Moby Dick* and the two random sequences consisting of shuffled and monkey texts is easy to show, even from the rank-frequency distribution. The limitation of the rank-frequency distribution is apparent only when words are considered separately. By considering texts in terms of sequences of more than one word, we can distinguish properly authored natural language texts from monkey texts.

For $n \geq 1$, the term *n*-gram indicates a length of n consecutive elements in a sequence X, denoted by $X_i^{i+n-1} \equiv X_i, X_{i+1}, \ldots, X_{i+n-1}$, for $1 \leq i \leq m - n + 1$. By convention, we call an *n*-gram a bigram when $n = 2$, and a trigram when $n = 3$. The case $n = 1$ implies simply words, which are often called unigrams to emphasize their use in comparison with bigrams and trigrams. Thus far, we have considered the rank-frequency distribution for unigrams, but it can also be generated for longer *n*-grams. For example, the bigrams for the very beginning of *Moby Dick* examined in Sect. 3.5 are $X_1^2 = $ "*Loomings .*", $X_2^3 = $ ". *Call*", $X_3^4 = $ "*Call me*", and so on. Similarly, $n = 4$ gives $X_1^4 = $ "*Loomings . Call me*", $X_2^5 = $ ". *Call me Ishmael*", etc. The same applies to character sequences: for example, for $n = 2$, $X_1^2 = $ "*Lo*", $X_2^3 = $ "*oo*", and $X_3^4 = $ "*om*", while for $n = 4$, $X_1^4 = $ "*Loom*" and $X_2^5 = $ "*oomi*", and so on.

Then, for all *n*-grams with a given n, the frequency is counted in a text, and the *n*-grams are ranked by a procedure similar to that described at the beginning of this chapter. The resulting rank-frequency relation can then be analyzed as a rank-frequency distribution. The left graph of Fig. 4.5 shows the rank-frequency distributions of *n*-grams for *Moby Dick* with different n. The plot for unigrams in red is exactly the same as the plot in Fig. 4.1. The remaining plots in blue and green are the rank-frequency distributions for $n = 2, 4, 8$. As can be seen, for *Moby Dick*, each plot for the different values of n very nearly forms a power distribution. Therefore, formula (4.1) applies not only to words but also to subsequences of length n, with smaller η for a larger n. The plots for $n = 1$ and $n = 2$ intersect each other,

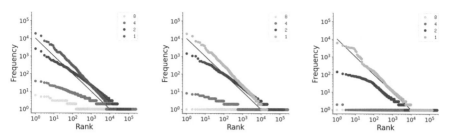

Fig. 4.5 Rank-frequency distributions of n-grams for *Moby Dick*, a shuffled text, and a monkey text. Each black line represents a power function with an exponent of -1

which is a common phenomenon in natural language texts. It is interesting to see that, even for $n = 8$, power-law-like behavior occurs with a certain small η.

The center and right graphs of Fig. 4.5 show the same results for the shuffled and monkey texts, respectively. For the shuffled text, because the sequence alignment is completely destroyed, the analysis using n-grams becomes meaningful for $n > 1$. The power-law tendency is visible for $n = 2$ and $n = 4$, similarly to the original text. At $n = 8$, however, η becomes almost 0 because of the effect of randomization. It is interesting to see that, even with the shuffling, the plot for bigrams is nearly the same as that of the original and crosses the unigram plot, although the slope is slightly lower (from the point of intersection, as compared with the black line). This result is not due to the randomization being insufficient in this sample. Rather, the power distribution raises the possibility of two frequent words occurring next to each other. For the shuffled text with $n = 4$, on the other hand, the difference is clear. As for the monkey text, as shown in the right graph, the differences from the original plots are also very clear. For $n = 2$, the plot does not show a power-function tendency, while the plots for $n = 4$ and $n = 8$ completely obscure each other, with $\eta = 0$.

These results show that n-grams follow power laws for natural language, even with a large n, but not for random sequences. A possible explanation of this feature is that there is a kind of memory underlying natural language text. In other words, many of the same n-grams appear in natural language texts even for a large n. Indeed, the left graph of Fig. 4.5 indicates that this effect is visible for natural language even with $n = 8$. As mentioned in Sect. 3.4, long memory is a characteristic of natural language that distinguishes it from random text. Part III will further verify this characteristic underlying a sequence of words, while the rest of Part II will examine the remaining characteristics of a population of unigrams.

4.5   Relative Rank-Frequency Distribution

Later in Part II, we will sometimes need to compare the rank-frequency distributions of different texts. As seen in Sect. 4.2, rank-frequency distributions can appear at different heights depending on the amount of data, although the overall tendency of

Fig. 4.6 Relative frequency
with respect to the rank in
order of the relative
frequency of words in the text
of *Moby Dick*. The black line
represents a power function
with an exponent of −1

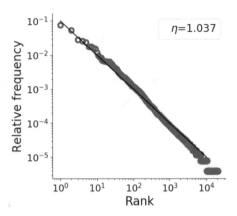

the population remains the same. One way to tackle this problem is to consider the
relative frequency instead of the absolute frequency.

Let the relative frequency of a word $w \in W$ be $z \equiv f/m$, where f is the
frequency of w and m is the total text length in words, as defined before. If formula
(4.1) holds, then

$$z = z_{\max} r^{-\eta}, \tag{4.2}$$

where $z_{\max}$ is the maximum relative frequency. The value of η is the same as in
formula (4.1), because m is only a constant factor dividing both sides of formula
(4.1) and therefore does not affect the exponent. Figure 4.6 shows the resulting
distribution for *Moby Dick*. The difference between Fig. 4.6 and the right graph of
Fig. 4.1 is the vertical axis, indicating the proportion of the frequency of words, z.

The black line in the figure is a function with slope −1 (*not* a fitting function).
The deviation of the red plot from this line is clear, and we are interested in the actual
value of η acquired from the plot. In comparison with the black line, all graphs thus
far for *Moby Dick* have shown a slightly steeper slope than −1. Obtaining the value
of η involves fitting the actual plot to a straight line. This process requires care,
especially for the case of a rank-frequency distribution, because the plot deviates
from a harmonic function, especially at the head and the tail. At the head, depending
on the special nature of highly frequent elements, the plot is often not straight. At the
tail, on the other hand, the plot appears as a step function for the discrete frequency
values.

Although the least-squares method, which minimizes the distances from the fit
line, is the most popular means of curve fitting, it has been reported to have problems
for data plotted on double-logarithmic axes. Clauset et al. (2009) and Gerlach and
Altmann (2013) have indicated better ways to solve this problem. In particular, we
will use the maximum-likelihood method for fit estimation. (See Chap. 21.1 for the
definition.)

In the case of *Moby Dick*, the fitted slope gives $\eta = 1.04$. The straightness of the plot can be partly measured in terms of the *goodness of fit*.[2] For comparison, η is exactly the same for the shuffled text, and 0.82 for the monkey text. Although the overall slope seems not much different from that of *Moby Dick*, regression by likelihood maximization gives this smaller value because of the heavy tail for the monkey text. Therefore, *Moby Dick* can be distinguished from monkey text even only by this value. Overall, the value of η shows a deviation from 1 of 0.04 for *Moby Dick*, which is small but requires attention. Chapter 5 discusses this variation in η.

---

[2]Chapter 21.1 defines measures for the goodness of fit, which depend on the fitting method. For the maximum-likelihood method, the goodness of fit is primarily measured by the value of the negative log-likelihood, LL. In the case of *Moby Dick*, $LL = 6.692$ for $\eta = 1.037$. At this point, we can only say that, for *Moby Dick*, the plot is pretty straight, and the fit is therefore good. Section 5.1 will compare these values across different datasets. The rest of the book gives measures indicating the goodness of fit in the footnotes.

Chapter 5
Bias in Rank-Frequency Relation

As shown at the end of the previous chapter, the rank-frequency relation of *Moby Dick* almost follows a power law with an η value close to 1. The goal of this chapter is to see how well Zipf's law holds among various kinds of texts and data. A text is typically written by a *single* author, but other corpora consist of collections (e.g., newspapers, collections of literary texts). Analyses like the one conducted here have also been reported beyond written texts, for speech (including infant utterances) and even program source code and music.

It is remarkable that Zipf, in his time, discovered power laws that hold for varied texts across languages, authors, and even beyond. By today's standards, however, he collected only a small amount of data; moreover, he did so without modern computers, which make it easy for us to analyze a large amount of data.

The overall Zipf property, with $\eta = 1$, must be considered approximate, and we thus need to consider the degree to which this approximation holds. Hence, this chapter summarizes the universality of and deviation from Zipf's law. It also shows the state-of the art understanding of the demography underlying language.

5.1 Literary Texts

The first issue is to discover how representative *Moby Dick* is of texts in general. To do so, we need a corpus of long texts suitable for statistical analysis. As *Moby Dick* is a literary text, we will compare it against a corpus built mainly from Project Gutenberg, a collection of literary masterpieces, and Aozora Bunko, a Japanese literary text archive.[1] Chapter 22, in particular Table 22.1, describes the details of this data. The texts cover 14 languages, but the majority are in English. They

---

[1]Project Gutenberg includes few Japanese texts, so more were separately acquired from Aozora Bunko.

© The Author(s) 2021
K. Tanaka-Ishii, *Statistical Universals of Language*, Mathematics in Mind,
https://doi.org/10.1007/978-3-030-59377-3_5

include many long masterpieces, such as *The Pickwick Papers* by Charles Dickens in English, *La Débâcle* by Emile Zola in French, *Phänomenologie des Geistes* by Georg Wilhelm Friedrich Hegel in German, and *Journey to the West* by Cheng'en Wu in Chinese. The corpus consists of 1142 long single-author texts[2] whose sizes in the archives are above a size threshold of 1 megabyte (including metadata). The texts were preprocessed to remove all metadata.

Following the argument presented in Sect. 4.5, the relative rank-frequency distribution was plotted on double-logarithmic axes for every text, and a straight line was fitted by the maximum-likelihood method. The resulting rank-frequency distributions were analyzed in two ways:

- the degree to which the plot follows a power law, and
- the value of η.

Note that if the plot largely deviates from following a power law, we cannot estimate η by fitting to a straight line, i.e., a power function.

For the collection of literary texts, in fact, the plots were aligned pretty straight on double-logarithmic axes, similarly to *Moby Dick*. The average goodness of fit was also similar to that of *Moby Dick*.[3] The results show that the rank-frequency distribution generally follows a power law fairly well for literary texts.

Thus, η could be acquired for every text. Figure 5.1 shows a histogram of the η values. They gather around a mean of 1.04, with a standard deviation of 0.021. Relatively few texts have an η value close to 1, with the majority having a larger

Fig. 5.1 Histogram of η values for 1142 long literary texts

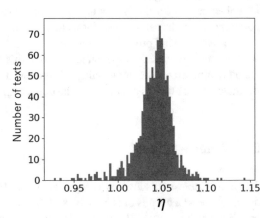

---

[2]Precisely, texts were included when they listed a single author. Even when a single author was listed, some texts, such as *The Traditional Text of the Holy Gospels* by John William Burgon, could have merely been collected by a single person, making it doubtful whether they are really *single authored*.

[3]Specifically, the average LL value was 6.57, with a standard deviation of 0.30. This mean value of LL is pretty similar to that reported for *Moby Dick* at the end of the previous section. As mentioned previously, the definition of LL appears in Sect. 21.1. LL becomes small when the fit is good.

value. The small standard deviation of 0.021 indicates that the literary texts overall have η slightly above 1, similar to that of *Moby Dick*.

The text with the steepest slope is *The Traditional Text of the Holy Gospels* in English by John William Burgon, with $\eta = 1.14$, and the second steepest is for *Nangoku Taiheiki* in Japanese by Sanjugo Naoki (a popular writer), with 1.12. On the other hand, the text with the shallowest slope is *Le maccheronee* in Italian by Teofilo Folengo (a poet), with $\eta = 0.93$, and the second shallowest is a Finnish translation of *Ivanhoe* by Walter Scott, with 0.94. Note that the versions of *Ivanhoe* in Dutch and English have η of 1.04, very near the mean.

The value of η being consistently close to 1 indicates that there must be some specific reason (or some underlying "theory") for that value, as Zipf suggested. Section 4.3 gave an argument in relation to a monkey text, and furthermore Chap. 15 will overview additional theories and show some random sequences that produce $\eta = 1$.

The standard deviation of 0.021 is small, but the mean value of 1.04 suggests that the 1142 literary texts have a slight bias in their vocabulary population, containing more frequent words and fewer rare words. Zipf reported that texts authored by people with schizophrenic disorders have steeper slopes (Zipf, 1949) [Chap. 7]. Given that longstanding folk wisdom associates genius with such disorders, the mean value of η being larger than 1 for the collection of literary texts could be due to that kind of reason, as Project Gutenberg is a large collection of literary masterpieces. Verification of the hypothesis would require analyzing texts written by authors with mental disorders; van Egmond (2018) reported a careful study on this topic. As such analysis would require a professional background in psychological disorders, we will assume here that the reason for the vocabulary population bias involves multiple factors.

Clarifying why the mean slope is steeper than -1 requires us to identify the factors that lead to a specific deviation from Zipf's law. Later, this chapter will consider cases of clear deviation from Zipf's law, and it will further examine the factors underlying the deviation of 0.04 mentioned above. But first, we will explore other phenomena, beyond texts, that are characterized by Zipf's law.

5.2 Speech, Music, Programs, and More

Apart from natural language texts and random sequences made from them, many other kinds of data approximately follow Zipf's law. Figure 5.2 shows the rank-frequency distributions for three examples taken from sources other than literary texts[4] (see the data listed in Sect. 22.3). The first graph is for speech data, specifically, a transcription of one of the longest TED talks in English. Speech data is usually smaller than the typical data used in this book, and this transcription is

---

[4] $LL = 5.817$ for TED, $LL = 6.070$ for Beethoven, and $LL = 7.155$ for Haskell.

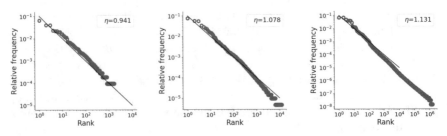

Fig. 5.2 Rank-frequency distributions for speech data (a TED talk), music (Beethoven's Symphony No. 9), and program source code (Haskell). In each graph, the black line has a slope of -1, and the fitted value of η is shown

about one-tenth the size of a literary text. The plot is pretty straight, with $\eta = 0.94$, which is smaller than 1 but within the range of values for literary texts described in the previous section. In fact, TED talks in various languages usually follow Zipf's law.

The second graph is for music, which was also considered in Zipf (1949) [Chap. 8]. Today, music data is more readily available, and the example here is Beethoven's Symphony No. 9, transcribed from MIDI data (cf. Sect. 22.3).[5] This data presents a slightly convex, power-like result, with an overall η value of 1.08. Music data tends to be short even for the longest symphonies. Twelve other long pieces of classical music were similarly tested, and they shared the global tendency of the Beethoven example. Some, however, showed disjunctions in the alignment of the plot or deviations similar to cases that the following section will examine.

The third graph is for computer program source code, examples of which were crawled and archived from the Internet. The example shown here was in the Haskell programming language. The η value is 1.13. Program source code examples in five different languages were examined, as described later in Sect. 22.3. The plots were similar, except for small disjunctions in their alignments.[6]

A rank-frequency distribution with a slope of about -1 is also fairly common for phenomena other than language, as Zipf indicated (Zipf, 1949). For example, plots of salary versus salary rank in an organization are known to follow a power law with a slope of -1 (Zipf, 1949) [Chap. 11]. Similarly, plots of city population versus population rank also have a slope of -1 [Chap. 10].

If such wide-ranging phenomena roughly follow Zipf's law, it would suggest that there is some underlying statistical mechanics. One possibility was introduced in the

---

[5]There is no conventional way to transform music to a sequence of words. Therefore, for this book, MIDI data that encodes a performance was transcribed to a text by a software application called SMF2MML. This produces a sequence of sounds in a text format. Every encoded note is considered a word in this analysis. There could be many other ways to analyze music, and this way is one possible crude starting point.

[6]The disjunctions were deemed to represent the differences in behavior among the different categories of words in source code, namely literals, operators, reserved words, and identifiers.

previous chapter via the behavior of a monkey text, and Chap. 15 will consider it further. In this chapter, we will examine deviations from a slope of -1 and their underlying factors. In the case of language, both the deviations and the nonlinearity of the plot reveal some interesting aspects of what language is.

5.3   Deviations from Power Law

The rest of this chapter analyzes how rank-frequency plots for language deviate from Zipf's law. As we have already seen, slight deviations can occur at the head and the tail of the distribution. For *Moby Dick*, $\eta > 1$, and the plot shows deviation at the head. This is a typical tendency for literary texts, because some of the most frequent functional words behave differently than in other texts. It is difficult, in fact, to find a completely straight example with a slope of -1. Such deviations were pointed out by Zipf himself. Thus, researchers have proposed to reformulate the fit function for the rank-frequency distribution. The earliest such proposal, by Benoit Mandelbrot, enlarged the number of parameters underlying the power law (Mandelbrot, 1953, 1965). More recently, Baayen (2001) [Chap. 3] investigated various possible alternative functions. At the same time, there have been reports of classes of deviations for some datasets that apparently do not show power-law behavior. The plots for these datasets are convex, starting with a slower drop at the head and dropping faster at the end. Where exactly this convex tendency appears— at the head, middle, or tail of the distribution—depends on the data kind.

Strongly biased rank-frequency distributions have been reported for at least three kinds of datasets: large-scale data, child-directed speech, and Chinese characters. Figure 5.3 includes an example of each. The axis scales of the three panels are normalized to enable better comparison of the plots' tendencies. For large-scale data, the bottom plot in the right panel shows the rank-frequency distribution of a large collection of *Wall Street Journal* articles. The head is close to power-law behavior, but the tail drops faster. The middle plot in the rightmost panel shows the result for "Thomas," the largest dataset in the CHILDES database of child-directed speech. The overall tendency is convex, with almost no straight part. Finally, the first plot in the middle panel shows the tendency for Chinese characters. The plot is only straight for the first third before dropping strongly from the middle toward the tail.

Zipf plotted rank-frequency distributions and analyzed the latter two cases. The abundant large corpora available today, however, serve to update and illuminate the deviations from power-law behavior and lead us to consider an alternative fit function; accordingly, we will have to reconsider the nature of language. The following subsections describe the consequences of these deviations found in three major datasets.

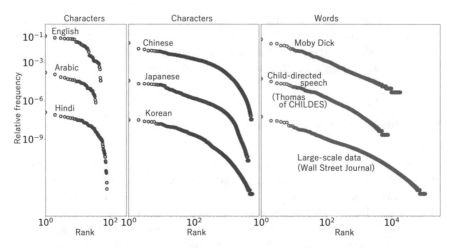

Fig. 5.3 Rank-frequency distributions for characters in six languages, shown in blue in the first two panels, and for words in *Moby Dick*, child-directed speech (the Thomas dataset in CHILDES), and large-scale data (*Wall Street Journal* articles), shown in red in the right panel. The left panel represents scripts with a relatively small number of characters (English, Arabic, and Hindi), while the middle panel represents scripts with a larger number of characters (Chinese, Japanese, and Korean). The right panel shows the distributions for datasets with typical large deviations from Zipf's law, with *Moby Dick* shown for comparison. To change the vertical placement for greater legibility in each panel, the frequencies for Arabic, Japanese, and Thomas were multiplied by 10^{-3}, and those for Hindi, Korean, and the *Wall Street Journal* were multiplied by 10^{-6}

5.3.1   Scale

The property of being scale free would require Zipf's law to hold for large-scale data. The reality shows that this is not the case, as clear biases have been reported to occur for larger-scale texts.

The last plot of Fig. 5.3 showed an example, for the *Wall Street Journal*. The data consisted of over 22 million words (see Sect. 22.2), which is a scale around 100 times larger than that of a typical literary text. The plot shows a clear tendency to drop faster than a straight line, and the slope from the middle to the tail differs from that of the head. Such a disjunctive change in the power exponent is called a *crossover*. Plots for large-scale language data typically shows a crossover around the middle.

Such tendencies were reported by Montemurro (2001), who examined a collection of Shakespeare's works and found that the rank-frequency distribution had a crossover. This implies that a crossover not only results from multiple authors but also can appear in a single-author case. To model this tendency, Montemurro (2001) introduced a Tsallis differential equation model and showed an alternative fit model.

Gerlach and Altmann (2013) considered multiple sets of large-scale data and found a tendency that the tail consistently drops. As a result, they sought the best two-parameter fit functions among various candidates and found that a double power law gave the best fit. A double power law is a function combining two power laws that are conjunct at some $r = a$, as follows:

$$f(r) \propto \frac{1}{r}, \qquad \text{if } r < a$$

$$a^{b-1}r^{-b} \qquad \text{otherwise,}$$

where a and b are functional parameters. A crossover suggests that two phenomena with inherently different exponents are mixed. In turn, this might suggest that a vocabulary consists of two different sets with different natures. Gerlach and Altmann (2013) considered the mechanism of the tail drop by introducing a hierarchy of generative models.

The tail-drop phenomenon suggests that the increase in vocabulary with respect to the text length decelerates when a text becomes sufficiently large. The problem of large-scale data reveals a common factor limiting the vocabulary growth with increasing text size, for both texts with one author and large collections of texts such as newspaper articles with many authors. Single-author texts are generated from one information source: the author.[7] Similarly, newspapers have a limited variety of content. As a result, the size of the vocabulary is not commensurate with the quantity of data in these cases.

Bernhardsson et al. (2009) studied the density function of a vocabulary population, as defined in the next chapter, and showed how the exponent of its power term decreases systematically with the text size. Their rationale for such change is that *we tend to repeat words more when writing a longer text*. To explain this change, they introduced the concept of a *metabook*, from which a text is sampled. Their concept explains how vocabulary growth slows down for larger and larger texts. Section 6.2 will consider the nature of vocabulary size growth again.

Reconsideration of the analysis of Bernhardsson et al. (2009) from a rank-frequency distribution perspective will reveal that η increases with respect to the text length.[8] Figure 5.4 shows the relation between the text length and η for the literary texts considered here. Every point corresponds to one text, with its estimated η plotted with respect to its length. The figure shows how η generally increases with the text length. The points are also distributed vertically, however, so the length is not only a factor of η. Nevertheless, the fact that Project Gutenberg texts used in this book are all long could be one reason why the mean η value is larger than 1 (i.e., 1.04).

---

[7]The vocabulary size for an individual has been analyzed in the cognitive linguistics field (Aitchison, 1987).

[8]Precisely, Bernhardsson et al. (2009) studied the density function of a vocabulary population, which takes a different perspective from that of a rank-frequency distribution. They reported that ζ in the density function, when defined as given in formula (6.3) in the next chapter, decreases as the text length increases. Theoretically, this corresponds roughly to the increase in η, as the next chapter will explain via formula (6.2). They showed this change by taking portions of literary texts.

Fig. 5.4 η with respect to the text length for 1142 literary texts. Each point represents a text. There is an increasing tendency, but it is clearly not the only factor affecting the distribution of η

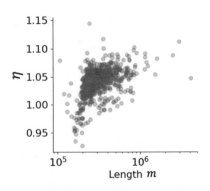

5.3.2   Speaker Maturity

Zipf plotted rank-frequency distributions for the speech of 5- to 7-year old children Zipf (1949) [Chap. 4]. The data, comprising around 10,000–25,000 words, showed that the children had already acquired rank-frequency distributions similar to those of adults.

Child-directed speech data has been archived in the CHILDES database of the TalkBank System led by Brian MacWhinney of Carnegie Mellon University. One of the largest datasets contains the utterances of Thomas (Lieven et al., 2009). They were recorded when he was 2- to 4-years old, younger than the children studied by Zipf (1949). The dataset amounts to nearly 450 thousand words after removal of adult utterances (by the mother and a tutor, which are labeled). This is around 20 times larger than the data used by Zipf.

The middle plot in the right panel of Fig. 5.3 shows the rank-frequency distribution of the Thomas dataset. It is characterized by a continuous, convex plot. Although this tendency is a little difficult to see because of the scale, a closer look reveals how Thomas's distribution is continuously nonlinear. The convex tendency is a common feature of the largest datasets in CHILDES. The plot is not only convex but also has a steeper slope for the overall tendency, although it is difficult to exactly define the slope of such a convex plot.

As mentioned by Zipf (1949) [Chap. 4], such a convex tendency may be due to echolalia. Very young children, especially, repeat the words of adults. For example, when asking a child, "Do you want a cookie?", the child often answers with "cookie" instead of "yes." The effect of this echolalia is to replace the most frequent words with some other content words, which are often located in the middle or the tail of a rank-frequency distribution. Especially for Thomas, given his age, the convex tendency can be partly explained by the phenomenon of echolalia. At the same time, children use more frequent words than adults do. This surely leads to a steeper slope for the rank-frequency distribution.[9] As a child grows, the convexity of the rank-frequency distribution straightens into a power function. Clarification of

---

[9]Baixeries et al. (2013) indicated how the overall η value decreases with age.

the cognitive reason underlying the convex shape of the rank-frequency distribution for young children will require more detailed future studies.

5.3.3 Characters vs. Words

Daniels and Bright (1996) classified writing scripts around the globe into six different categories: alphabet, abjad (consonantal writing, as in Hebrew and Arabic), abugida (block characters, typically represented by Hindi and Thai), featural writing (Korean only), syllabary (Japanese kana script), and logosyllabary (Chinese). The categorization shows a rough continuity from sound to meaning—from phonemes (alphabetic, consonantal) to morphemes (logosyllabic). As noted by Coulmas (1996), however, it is often difficult to completely delineate the categorization.

The first and second panels of Fig. 5.3 show the rank-frequency distributions for languages representing each of the six categories: English, Arabic, Hindi, Chinese, Japanese, and Korean. Each dataset was acquired from publicly available corpora.[10] All the plots in the first panel show some convex tendency,[11] whereas those in the second panel are straight from the head to the middle. We can see a straightening tendency going from the left panel to the right panel. In other words, the plots tend to *powerize* from left to right.

These characteristics for different languages have been studied previously. Good (1953) and Bell et al. (1990) reported the drop in the distribution for alphabetic scripts, while Lü et al. (2013), Deng et al. (2014), Allahverdyan et al. (2013), and Nabeshima and Gunji (2004) found that the rank-frequency distribution has a fast drop at the tail for the Chinese and Japanese scripts. Each of these reports suggested its own fit function: for example, Nabeshima and Gunji (2004) suggested replacing the power function with a Weibull function for the Japanese language. Each proposed function was only for the rank-frequency distribution of the target language; for other languages, the fit was not studied.

Note that alphabetic scripts are phonetic, whereas words are semantic. Somewhere in between lie Chinese characters. The semantic aspect of Chinese characters is usually emphasized in comparison with alphabets, but they also have phonetic aspects, such as an ideographic part contributing a consistent pronunciation across characters that commonly include that part. Furthermore, the syllables of syllabic scripts such as Hindi often have some relation with semantics. Because the straightening of the rank-frequency distribution progresses between phonetics and

---

[10] As explained in Sect. 22.4, the results presented for characters in the first two panels of Fig. 5.3 exclude all symbols, such as spaces, punctuation marks, and so on. The analysis was conducted only on the set of characters for each language, because there is a clear, available definition of the set of characters for each script, based on Unicode.

[11] Because an exponential function presents a convex tendency, such plots are often roughly called "exponential." The rank-frequency plots of English and Arabic do present a sort of rough linear tendency on semi-log axes, but with some disjunctions among the points. Whether these plots are really exponential is an issue that will require future work.

semantics, it might seem reasonable to think that the mechanism underlying the straightening relates to meaning.

This possibility of powerization in relation to *meaning*, however, must be carefully considered. Recall that the monkey text examined in Sect. 4.3 had a rank-frequency distribution following a power function. The distribution of the monkey text for *characters* should be similar to the top plot in the left panel of Fig. 5.3. Given that a monkey text, whether in terms of words or characters, has no meaning at all, the straightening should be purely statistical. Rather, the contrast reflects that characters constitute a system with a finite number of elements, whereas words constitute a system with a large number of elements, perhaps infinite, acquired by aligning a finite number of elements. When the number of elements is limited, there is a drop at the tail, but when the number increases, the demography changes and gives a power law as a statistical consequence.

It is not that obvious, therefore, whether meaning underlies powerization. Rather, the increase in the number of elements is exploited in a language system to represent meaning. Obviously, we need many thousands of specific signs to represent different objects existing in reality. The representation could perhaps be constructive, using multiple sounds in combination. Such a way of constructing words through combination would perhaps have an effect similar to a monkey text, as seen in the previous chapter, thus leading to the production of Zipf's law. These statistical changes in the demography with an increasing number of elements would provide the necessary framework—a powerized system—to constitute language.

5.4 Nature of Deviations

The above deviations are statistical for the case of characters, but they have cognitive origins for the case of child-directed speech. The deviations in large-scale data, on the other hand, have a number of causes, requiring cognitive modeling of the words in mind and the statistical consequence. Note that the three factors analyzed in the previous section—scale, maturity, and the number of elements—are all present in any text, to some extent. The overall rank-frequency distribution of a text is a mixture of these different factors at different degrees, together with other possible factors causing other kinds of deviations.

The deviations from a power law indicate a limit on the degree of self-similarity underlying language. As mentioned in Sect. 3.1.3, a language sample is finite, so the degree of self-similarity becomes limited at some scale. Given that language is used by humans and involves both statistical and cognitive deviations, the causes of the deviations should be carefully separated to contemplate the true nature of language.

Overall, Zipf's law is no more than a rough first approximation. Indeed, many great minds have suggested alternatives to Zipf's law. Formulation of a good alternative model will require other perspectives to analyze the population of a text. The next chapter introduces two such perspectives by changing how we view the population.

Chapter 6
Related Statistical Universals

This last chapter of Part II further considers the nature of a vocabulary population in terms of two related properties that have a mathematical relation with Zipf's law: the density function and the vocabulary growth. Similarly to Zipf's law, both nearly indicate power-law behavior but are subject to some deviations.

6.1 Density Function

The rank-frequency distribution is one way to examine the vocabulary population, but it is not a usual way of conducting a statistical analysis. A more common statistical approach examines the histogram of words of certain frequencies, f, giving a distribution called the density function. It is based on counts of the number of words of the *same frequency*. In the case of *Moby Dick*, for example, the two words "yet" and "still" occur 299 times, while seven words—"wind," "rather," " often," "name," "myself," "morning," and "four"—occur 68 times. Obviously, the smaller f is, the more words of that frequency occur. The function, defined as the number of words occurring f times, is denoted as $g(f)$; thus, for *Moby Dick*, $g(299) = 2$ and $g(68) = 7$. Because this function g is difficult to compare across texts, it can be normalized as in Sect. 4.5. Hence, an equivalent function $P(f)$, the *density function*, is introduced to give the relative number of words with frequency f, i.e., $P(f) \approx g(f)/v$.[1] For example, *Moby Dick* has $v = 20,472$, so $P(299) = 2/v = 0.000098$, and $P(68) = 7/v = 0.00034$.

The original version of this chapter was revised. The correction to this chapter is available at https://doi.org/10.1007/978-3-030-59377-3_23

[1] More precisely, the density function represents the distribution that defines the information source of the author of *Moby Dick*, Herman Melville. Such a true function is never known, however, so a guess is deduced from the word counts, as mentioned as P in the main text.

© The Author(s) 2021, corrected publication 2022
K. Tanaka-Ishii, *Statistical Universals of Language*, Mathematics in Mind,
https://doi.org/10.1007/978-3-030-59377-3_6

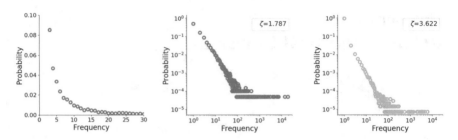

Fig. 6.1 Density functions of *Moby Dick* on normal axes (left) and double-logarithmic axes (middle), together with the density function of a monkey text (right) as defined in Sect. 3.5

Plotting $P(f)$ for *Moby Dick* gives the left graph of Fig. 6.1, which only shows part of the whole picture. Here, the least frequent words appear on the left, which is reversed from the case of a rank-frequency distribution. There is a decreasing tendency similar to that seen in the left graph of Fig. 4.6. Therefore, the plot for the density function is shown in green here to distinguish it from a rank-frequency distribution (shown in red in this book).

Plotting this result on double-logarithmic axes gives the middle graph of Fig. 6.1, which shows the overall picture. The density function tends to show a power-law decay, similarly to a rank-frequency distribution. The plot is more or less straight, but at the tail, the points become scattered. Nevertheless, as a rough first approximation, the density function of a text can be considered to globally follow a power law with a slope of $-\zeta$:

$$P(f) \propto f^{-\zeta}, \qquad \zeta > 0. \qquad (6.1)$$

Like the rank-frequency distribution, this power law underlying the density function also indicates how the vocabulary is scale free, implying that a text has the same vocabulary distribution, regardless of size. For *Moby Dick*, ζ was estimated as 1.79 by the maximum-likelihood method,[2] as shown in the upper right corner of the graph.

Figure 6.2 explains the relation between the rank and the density function. Consider the functional space of $g(f)$, which decreases as the frequency f increases. Here, neither axis is logarithmic, and the vertical axis indicates the number of words of frequency f. The rank r then corresponds to all the words with frequencies between f and f_{max}, i.e., the green area in the figure. Therefore, roughly speaking, because r is some integrated term of $g(f)$, if $g(f)$ is a power function, then $f(r)$ is also a power function,[3] and vice versa.

By considering the relation between, ζ and η (from Zipf's law, formula (4.1)), the following relation can be acquired:

$$\zeta = 1 + \frac{1}{\eta}. \qquad (6.2)$$

---

[2]$LL = 2.092$.

[3]Precisely, this depends on the value of ζ. For the values of ζ acquired from texts, the result of the integral becomes a power function.

Fig. 6.2 Schematic
illustration of the relation
among the rank r, frequency
f, and count $g(f)$

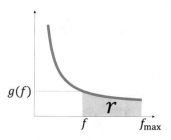

Section 21.3 gives a proof.[4] This relation suggests that when $\eta \approx 1$, then $\zeta \approx 2$.
Moreover, when η increases, ζ decreases, which is obvious, because r is the integral
over the colored region of Fig. 6.2.

For *Moby Dick*, which has $\eta = 1.04$, the theoretical value from the above formula
would be $\zeta^* = 1.96$, whereas the actual estimated value[5] by formula (6.1) is $\zeta = 1.79$, giving a deviation of about 0.17. Similar comparisons can be done for every
literary text mentioned in Sect. 5.1. From fitting the density function of every text,
the mean of ζ was 1.71, with a standard deviation of 0.11. The previous chapter
indicated that the mean value of η was 1.04, so the above formula gives $\zeta^* = 1.96$.
The resulting difference of 0.25 shows the limitation of the theoretical relation of ζ
and η, in addition to the limitation of fitting points scattered at the tail to a power
function, as seen in Fig. 6.1.

As for the behavior of a random text, the density function for random sequences
of shuffled text is exactly the same as that for *Moby Dick* itself, because the
frequency of each word in the shuffled text is exactly the same as in the original
text. On the other hand, the right graph of Fig. 6.1 shows the result for a monkey
text. The plot indeed looks quite different from that of the original *Moby Dick* in the
left graph. The ζ value is much larger, estimated as 3.62 from the yellow plot.[6] At
the end of Chap. 4, we found that a monkey text had $\eta = 0.82$. From that η value,
we can obtain a theoretical value of $\zeta^* = 1 + \frac{1}{0.82} = 2.21$, which shows a large
difference from the ζ value of *Moby Dick* estimated here.

This result clearly indicates that the monkey text has many more rare words than
Moby Dick does. In Sect. 4.3, we saw this difference between the rank-frequency
distributions of *Moby Dick* and the monkey text: the tail of the rank-frequency
distribution was longer for the monkey text. The value of $\zeta = 3.62$ here is not even
within a standard deviation of the mean ζ value in the histogram for literary texts,
and therefore, the density function better distinguishes this random sequence from a
real text, in comparison with the rank-frequency distribution. The difference is due
to the fact that the rank-frequency approach generates a point per word, whereas the

---

[4]The proof is based on a very simple, elegant consideration by Lü et al. (2010).
[5]$LL = 2.093$.
[6]$LL = 0.432$.

density approach generates a point for multiple words. The density function thus might seem to better capture the difference in the proportion of rare words.

For some other analyses, however, the rank-frequency distribution has an advantage over the density function. The former often shows more detailed information for analyzing individual words as compared with the latter. First, the density function is difficult to analyze for scripts with only a limited number of characters, such as an alphabet or an abjad. Second, the rank-frequency distribution often shows a clearer difference in shape between real texts, as compared to the density function. For example, for the different biases explained in Sect. 5.3, the slope of the density function is usually smaller than that for literary texts ($\zeta = 1.47, 1.49$, and 1.10 for the *Wall Street Journal*, Thomas, and Chinese characters, respectively).[7] This should be so, as these texts have fewer rare words, which leads to a smaller ζ. The systematic difference seen in the shape of the rank-frequency distribution, however, cannot be seen clearly with the density function.

Nevertheless, further study will require us to carefully examine the data points, especially those accumulated around the tail. One starting point is to apply the following function:

$$P(f) \propto f^{-\zeta} \exp(bf), \qquad (6.3)$$

as suggested by Bernhardsson et al. (2009). This formula quantifies the deviation at the tail in terms of the parameter b via $\exp(bf)$. As Bernhardsson et al. (2009) noted, this function generally fits well to density functions. Such representations enable finer analyses of the properties of language, as shown by the example of the relation between η and m in Sect. 5.3.1. We will return to this function in the following chapter when we examine the return distribution, as well.

Overall, studies of the distribution of a vocabulary population have usually attempted only one kind of statistical analysis. Yet there are multiple ways to examine the vocabulary population, including the rank-frequency distribution, the density function, and the type-token relation, which will be introduced in the next section.

6.2 Vocabulary Growth

The second property that is mathematically related to Zipf's law is described by Heaps' law (Heaps, 1978). It was reported before Heaps (Herdan, 1964; Guiraud, 1954), but by convention, this power property is called Heaps' law. It indicates the growth speed of the vocabulary size with respect to the text size. As mentioned in Sect. 3.1.1, the vocabulary size is said to be the number of *types* (i.e., $|W|$), while

---

[7]$LL = 3.376$, $LL = 3.180$, and $LL = 7.718$ for the *Wall Street Journal*, Thomas, and Chinese characters, respectively.

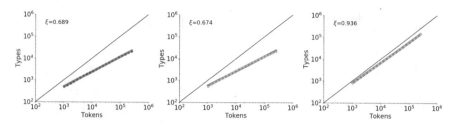

Fig. 6.3 Vocabulary growth plotted on double-logarithmic axes for *Moby Dick*, its shuffled text, and a monkey text, as defined in Sect. 3.5. The gray thick lines are fitted, whereas the black lines show $y = x$

the text size m is said to be the number of *tokens*, and therefore, the relation is often called a *type-token relation*, or vocabulary growth.

For Heaps' law, the vocabulary size v is plotted (vertical axis) with respect to the text size m (horizontal axis). This is done by taking the first $m \propto a^k$ words, where $k = 1, 2, 3, 4, \ldots$, for some value[8] of a. The left graph of Fig. 6.3 shows the type-token relation for *Moby Dick* on double-logarithmic axes.[9] The red points represent the data. The plot is rather straight, so the relation between v and m is described again by a power function, with slope ξ:

$$v \propto m^{\xi}, \qquad 0 < \xi \le 1. \qquad (6.4)$$

As shown in the upper left corner of the graph, the least-squares method estimates ξ as 0.69 for *Moby Dick*.[10] The thick gray line shows the fitted slope. In comparison with the black line with a slope of 1, i.e., $y = x$, (or, precisely, $v = m$), the exponent is clearly much smaller than 1. Note that this relation formulates the vocabulary growth in a scale-free manner, so the vocabulary growth does not depend on the text size. Such vocabulary growth is shared universally among language samples and is one of the statistical universals. Although here it remains within the size of *Moby Dick*, this growth indicates that the vocabulary size can increase infinitely, implying the open nature of the vocabulary that underlies language.

---

[8]$a = 1.3$ for the type-token relations shown in this book.

[9]For fitting Fig. 6.3, the least-squares method was applied (cf. Sect. 21.1). As defined in Sect. 21.1, the goodness of fit was measured by the residual ε. For *Moby Dick*, the shuffled text, and the monkey text, respectively, the error $\varepsilon = 105.322, 175.808, 45.538$. The values are very large because of the ranges of the axes. The comparison of residuals at least shows how the monkey text fits well to the fit line.

[10]$LL = 2.092$.

Fig. 6.4 Relation between η of Zipf's law and ξ for vocabulary growth for the collection of 1142 literary texts

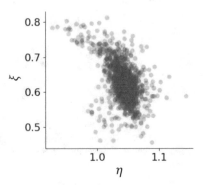

Many reports have considered the relation between η and ξ.[11] A number of studies (Lü et al., 2010; Baeza-Yates and Navarro, 2000; Serrano et al., 2009) have given the following relation:

$$\xi = 1/\eta, \qquad \eta > 1$$
$$1, \qquad \eta < 1. \qquad (6.5)$$

A proof is given in Sect. 21.4. As mentioned in Sect. 5.1, η is distributed around 1, and many texts have values below 1.

Calculation using the value of $\eta = 1.04$ for *Moby Dick* and the above relation gives a theoretical slope of $\xi^* = 1/\eta = 0.96$, which is rather different from the actual slope of $\xi = 0.69$ mentioned above. Nevertheless, the relation between ξ and η captures an important experimental tendency. Figure 6.4 plots the actual relation between η and ξ for the literary texts examined in Sect. 5.1. For every literary text, its estimated value of ξ is plotted with respect to η. For $\eta > 1$, ξ correlates in a decreasing manner with respect to η. Moreover, for $\eta < 1$, the points become more scattered around similar values of ζ near 0.8. Therefore, the theoretical relation seems to capture the rough global trend well from the disjunctive distribution of the points before and after $\eta = 1$.

The middle and right graphs of Fig. 6.3 show the results for the random sequences of shuffled and monkey texts, respectively. For the shuffled text (middle), the difference from *Moby Dick* is only minor, with consistent growth. The endpoints of the plots are the same, indicating v with respect to m for *Moby Dick*.

For the monkey text, however, the vocabulary grows much faster than for the original text, and this fits the theory much better. As mentioned before, a monkey text had $\eta = 0.82$. Here, as shown in the upper left corner of the right graph in Fig. 6.3, we have $\xi = 0.94$ for the monkey text. We can then compute a theoretical

---

[11] Among these are Baeza-Yates and Navarro (2000); Montemurro and Zanette (2002); van Leijenhorst and van der Weide (2005); Zanette and Montemurro (2005); Serrano et al. (2009); Lü et al. (2010).

value of η as $\eta^* = 1$, confirming that the theory fits well in this random case. This fast vocabulary growth is consistent with the rank-frequency distribution and density function results that we have seen. The value of ξ for a natural language text is typically 0.7 or even smaller. In contrast, a value close to 1 is evidence that a text is not written in natural language. In other words, the ξ value can distinguish a monkey text from the original *Moby Dick*. We will return to this fact in Part V.

The vocabulary growth plots for real texts also deviate from a power function. In general, type-token relations often show small global fluctuations around a power function, reflecting changes of context. Furthermore, when the rank-frequency distribution drops faster at the tail (Sect. 5.3), the vocabulary growth slows down. This phenomenon might not be apparent, however, at the size of literary texts, like in the left graph of Fig. 6.3. In addition to this corpus-dependent nature, Font-Clos and Corral (2015) showed that vocabulary growth slows down with respect to m when Zipf's law holds completely. Therefore, the decrease in growth for a large text has a statistical effect. Lü et al. (2010, 2013) thoroughly analyzed how actual plots deviated from their theory, as well.

The previous two chapters indicated that a population's deviation from a power function involves multiple factors. Thus, to reveal the nature of the vocabulary population, we should incorporate the different perspectives offered by the density function and the vocabulary growth, in addition to the rank-frequency distribution. The next chapter adds another perspective: that of the return interval distribution.

Part III
Property of Sequences

Chapter 7
Returns

Part II investigated the population of words, but the book thus far has not considered the properties underlying a sequence of words. Language forms a sequence, which characterizes what language is. Indeed, Sect. 4.4 showed that, for n-grams, the subsequences of natural language texts present a different nature from those of random sequences, even from the population viewpoint. This provides evidence that language has a kind of memory, meaning that a word in one part of a text influences words in other parts of the text.

As mentioned in Sect. 3.4, this book refers to this property of influence at long distances as *long memory*. Hence, Part III examines long memory from the perspective of statistical universals. A theme of the book thus far has been to discover how the nature of a natural language text differs from that of a random text. In Part II, we found that a monkey text indeed has different properties from those of its original text, especially in Chap. 6, even though it can produce a rank-frequency distribution similar to that of a natural language text. In contrast, the results in Part II for a shuffled text were almost or exactly the same as those for the original text, except in Fig. 4.5. Hence, a theme of Part III will be to distinguish a shuffled text from the original text.

As the first step in examining long memory, this chapter considers the behaviors of intervals between words. The following two chapters will examine it by using the long-range correlation and fluctuation analyses. Finally, the last chapter of Part III will show the complexity of linguistic sequences, which provides another way to analyze language from a point of view including both the population and long memory.

© The Author(s) 2021
K. Tanaka-Ishii, *Statistical Universals of Language*, Mathematics in Mind,
https://doi.org/10.1007/978-3-030-59377-3_7

7.1   Word Returns

A word w appears at some point in a text for the first time. Other words may succeed w before it appears again. This sequence continues to repeat, so it can be characterized as w, (possibly) other words, w again, other words, w again, and so on. These reappearances of a particular word are called its *returns*. One way to quantify a return is by the number of words between occurrences of w, called the *return interval length*.

For example, consider the word "Romeo" in the famous quote "Oh Romeo Romeo wherefore art thou Romeo" in the top half of Fig. 7.1. The thick vertical bars under the word sequence indicate the locations of "Romeo." The second "Romeo" is the next word after the first "Romeo"; we call this a one-word *interval* between the first and second occurrences. The third "Romeo" is the fourth word after the second "Romeo." Therefore, the interval sequence Q of "Romeo" in the quote is $Q = [1, 4]$. Doing this for all of *Romeo and Juliet* would generate an *interval sequence* for the word "Romeo" over the entire text. As this example indicates, an interval sequence can be acquired for every word in W, the set of all different words (types) in a text, and the distributions of the returns and interval sequences can be analyzed.

More generally, for a given word sequence $X = X_1, X_2, \ldots, X_i, \ldots, X_m$, the *interval* between X_i and X_j, where $1 \leq i < j \leq m$, is defined as $j - i$. Given a word $w \in W$ appearing in X, its return interval sequence is

$$Q_w = Q_1, Q_2, Q_3 \ldots, Q_i, \ldots, Q_{\#w-1}. \tag{7.1}$$

Here, $\#w$ indicates the number of occurrences of w in X. Q_i is the interval in the original sequence X between the ith and $(i + 1)$-th occurrences of w, for $1 \leq i \leq \#w - 1$; thus, $Q_i = q$, a positive integer. Therefore, Q_1 indicates the first interval, between the first and second occurrences of w; Q_2 indicates the second interval, between the second and third occurrences of w; and so on. The number of intervals for w is $\#w - 1$, which is thus the length of the sequence Q_w, as well. An interval

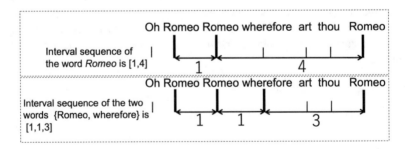

Fig. 7.1 Toy example of return analysis and the concept of extreme-value analysis applied to an interval sequence

sequence can be acquired for every word in W, except for hapax legomena. The next chapter will analyze the sequence Q_w, while this chapter analyzes the distribution of Q_i values.

Zipf himself (Zipf, 1949) [Chap. 2] used returns to analyze texts, although his analysis only considered the population of return intervals. The methods presented in the following chapters derive from more recent research reported in Ebeling and Neiman (1995), Altmann et al. (2009), and Tanaka-Ishii and Bunde (2016), each of which presented a new way to analyze return intervals. This chapter especially focuses on the methods of Altmann et al. (2009) and Tanaka-Ishii and Bunde (2016).

7.2   Distribution of Return Interval Lengths

The left graph of Fig. 7.2 shows a histogram on double-logarithmic axes of the return intervals for all words in *Moby Dick*. The horizontal axis indicates the return interval length, and the vertical axis indicates that length's frequency among the intervals for all words.

The main part of the left plot vaguely follows the power function indicated by the black line, whose exponent is -1. Here, a return distribution following a power function with an exponent of -1 might seem natural for a population following Zipf's law. Specifically, Zipf's law implies that the most frequent word occurs k times as often as the kth most frequent word does. Then, if words consistently occur with the same intervals, the kth most frequent word's interval should be k times longer than that of the most frequent word. This would mean that k-times longer intervals appear one-kth less frequently, giving the equation $y = 1/x$, the harmonic distribution.[1]

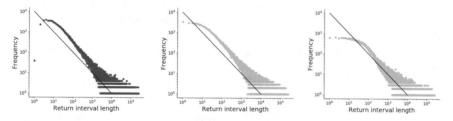

Fig. 7.2 Frequency of return interval lengths for words in *Moby Dick* (left), a shuffled text (middle), and a monkey text (right). The black lines represent a power function with an exponent of -1

---

[1]Many continuous natural phenomena commonly have a spectrum showing $1/f$ noise. Here, $1/f$ noise is defined as the frequency spectrum of a continuous time series following the function $y = 1/x$. This book is about a discrete sequence, however, whereas $1/f$ noise concerns a continuous time series. A clear relation between them would require defining what a "spectrum" signifies in a discrete sequence. There is an interesting work on this question for DNA (Voss, 1992).

The above argument roughly applies only to the mean behavior; in reality, return intervals fluctuate. The graph for *Moby Dick* in Fig. 7.2 deviates from the black line especially at the head, for $1 \leq q < 10$. In natural language, $q = 1$ indicates a case like the first repetition of *Romeo*. The same word following itself is a rare occurrence in natural language texts of the kind analyzed in this book.[2] Aside from the case of $q = 1$, the graph shows that English avoids repetitions with small return intervals. Once q exceeds 10, words repeat more frequently, and the histogram follows the black line more closely. The tendency to avoid close repetition is one piece of evidence for the dependence among words that underlies language.

Comparing the left graph with the middle graph, which shows the same result for a shuffled text of *Moby Dick*, can give us a better idea of how returns characterize natural language. Here, the mean intervals of *Moby Dick* and the shuffled text are the same, whereas the plot for the shuffled text is smoother. This is because the shuffled text has no dependence among words. Words *do not avoid* instantly repeating, so the interval lengths for $q < 10$ are naturally frequent, following the distribution of the whole population. The difference between the left graph and the middle graph with *no memory* should characterize natural language. Apart from the head, however, it is hard to distinguish the two graphs from only their appearances: the left graph seems slightly straighter than the middle one. The comparison thus points to the need for a better analysis method than just using the simplest population of return intervals.

The right graph shows the distribution for a monkey text, which is rather different. Recall that the population of a monkey text includes fewer frequent words and more rare words, so the distribution appears biased toward the tail. In Part II, the characteristics of a monkey text also differed from those of the original text, especially in Chap. 6. In contrast, the results in Part II for a shuffled text were almost or exactly the same as those for the original text, except in Fig. 4.5. Hence, for some time we will focus on how to distinguish a shuffled text from the original text.

To this end, we could analyze the behavior of rare words, as they occur in a different manner in a natural language text than in a shuffled text. However, the analysis would be challenging because rare words rarely occur. Figure 7.2 shows thick clouds of points at the tails, mainly indicating long returns, in all three histograms. To delve into these clouds, we will need to change how we regard the distribution of returns.

In particular, while the vertical axis in Fig. 7.2 represents the absolute *counts* of the intervals, they can also be treated proportionally, as was done in the previous chapter. In this case, $P(q)$ denotes the probability distribution of an interval length q, approximated by the relative frequency.[3] The following sections attempt to

---

[2] Repetitions in language within a close range of sounds or words are studied in the field of linguistics and referred to by the term *reduplication*. For an introduction, see Rebino (2021). Texts with reduplication would appear different if they were statistically analyzed as mentioned here.

[3] In this chapter, there are two cases for considering the relative frequency: with respect to a specific interval sequence Q_w for a word w, or with respect to *all* intervals. In the former case, $P(q) = \#q/|Q_w|$, where $\#q$ denotes the number of occurrences of intervals of length q. In the latter case, for every type of word, each occurrence after the first one can be associated with an interval; thus,

differentiate *Moby Dick* from a shuffled text by using various quantities derived from $P(q)$.

7.3   Exceedance Probability

First, instead of using $P(q)$, we can use the *exceedance probability* function $S(q)$, i.e., the probability that an interval length exceeds q. $S(q)$ thus gives the proportion of intervals longer than q. Figure 7.3 schematically illustrates this concept. When $q = 0$, $S(q) = 1$, because any return interval has a length greater than 0. With increasing q, the probability $S(q)$ decreases, and $S(q) = 0$ at $q = q_{max}$.

The exceedance probability can be defined as

$$S(q) \equiv 1 - Cum(q), \tag{7.2}$$

where $Cum(q)$ is the cumulative distribution function of $P(q)$,[4] that is, $Cum(q) - Cum(q - 1) = P(q)$. $Cum(q)$ starts at 0 and reaches 1 as q increases to q_{max}; it is thus troublesome to plot on a logarithmic axis, because the logarithms would start from $-\infty$ (at $Cum(q) = 0$). In contrast, $S(q)$ drops from 1; therefore, on a

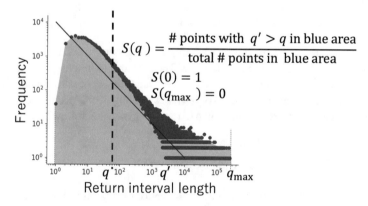

Fig. 7.3 Schematic figure illustrating the exceedance probability

---

the accumulated frequency equals $m - v$. Therefore, $P(q)$ can be estimated as a relative frequency, i.e., $P(q) = \#q/(m - v)$. Section 7.3 and Fig. 7.7 consider the case of all intervals, whereas the rest of the chapter considers the case of a specific interval sequence Q_w.

[4]The cumulative distribution function $Cum(q)$ of $P(q)$ is defined as follows:

$$Cum(q) \equiv \int_0^q P(q)dq. \tag{7.3}$$

logarithmic axis the value decreases from $\log 1 = 0$, enabling better comparison across different texts.

The exceedance probability by itself, however, does not distinguish *Moby Dick* from a shuffled text. This is natural, because the intervals are inclusive of all return values for both texts. We will see the shape of $S(q)$ for all words at the end of the chapter, but first we will distinguish *Moby Dick* from a shuffled text by analyzing $S(q)$ for the returns of *individual* words.

7.4 Bias Underlying Return Intervals

In a shuffled text, a word occurs independently of other occurrences. As explained in Sect. 3.5, a shuffled word sequence follows an (almost) i.i.d. process, so word occurrences are characterized only by the word population. The sequence thus has the same stochastic characteristic at any point in the sequence.

Let us consider a specific return interval sequence Q_w for a word w. The probability $P(q)$ and the exceedance probability $S(q)$ are considered via the relative counts of q in Q_w. In this case, the return intervals follow an exponential distribution. This property is described as follows, with $P(q)$ being the probability of interval length q occurring:

$$P(q) \propto \exp(-q/\lambda), \tag{7.4}$$

where λ is the mean return interval length, i.e., $\lambda \equiv m/f$. Section 21.5 explains why formula (7.4) holds for a shuffled text. The probability of a large return interval is exponentially small. In other words, longer intervals become exponentially less likely to appear, indicating a lack of *memory* among word occurrences. In this exponential case, the condition that $S(0) = 1$ gives the following exceedance probability:

$$S(q) = \exp(-q/\lambda). \tag{7.5}$$

Altmann et al. (2009) showed that the behavior of a word in a natural language text deviates from an exponential function and is characterized instead by a stretched exponential function:

$$S(q) = \exp(-\kappa(q/\lambda)^{\theta}), \tag{7.6}$$

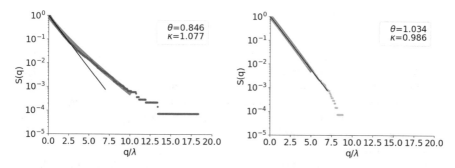

Fig. 7.4 Exceedance probability $S(q)$ for the word "the" in *Moby Dick* (left) and in a shuffled text (right). The black lines have a slope of -1, and the thick gray lines represent the fitted stretched exponential function

where κ and θ are functional parameters. This corresponds to the Weibull distribution, which was also mentioned in Part II. The main difference from formula (7.5) is the exponent θ applied to q.[5]

$S(q)$ therefore clarifies the behavior of a word in a text. For $S(q)$ to be useful, however, the word should be a frequent one, so that a sufficient number of intervals can be analyzed. As an example, consider the word "the," which is certainly a very frequent word in English. Figure 7.4 shows $S(q)$ with respect to q/λ for *Moby Dick* and a shuffled text. Given the functional forms of $S(q)$ in formula (7.5) and (7.6), only the y axis is logarithmic. For the shuffled text, the plot should follow the straight line $y = -x$, so that the intervals follow an exponential distribution, and this is indeed the case in the figure. On the other hand, for *Moby Dick* the tail is almost entirely above the black line. Comparing the two figures shows that the behavior of "the" clearly differs between the original and shuffled texts. *Moby Dick* fits the stretched exponential function well, with $\theta = 0.85$ and $\kappa = 1.08$.[6]

The behavior of occurrences of the word "the" can thus be distinguished between *Moby Dick* and the shuffled text. It is somewhat surprising that the exceedance probability of "the" deviates from an exponential function, as we would not expect very long intervals between occurrences of "the." The result suggests, however, that large intervals are more probable than in the shuffled text. Moreover, given that "the" shows such specific behavior, the behaviors of rare words should be even more interesting.

---

[5]Altmann et al. (2009) also showed that a renewal process produces this stretched exponential function. It is not obvious, however, how to integrate the population characteristic seen in Part II into the renewal process to form a sequence.

[6]The points here were fitted by the least-squares method (cf. Sect. 21.1), with $\varepsilon = 0.0226$ for *Moby Dick* and $\varepsilon = 0.00929$ for the shuffled text.

7.5 Rare Words as a Set

It is difficult to apply the same analysis to rare words, as many are hapax legomena. The length of the word sequence Q_w might be too short to effectively analyze the distribution of rare words. While Part II considered words equally, no matter what their frequencies were, this part analyzes how words occur in a sequence. For infrequent words, the statistical analysis will have limited value unless we have an effective way to analyze the rarity.

An effective way to approach this problem is to treat many rare words as instances of a single special word (Tanaka-Ishii and Bunde, 2016). In particular, we can group a quantity of $1/\psi$ of rare-word tokens into a set W_ψ as if they were a single word. Figure 7.5 schematically shows the concept via a rank-frequency plot of *Moby Dick*. The tail includes the least frequent words. The word types with the lowest ranks are put in W_ψ, as indicated by the gray zone in the figure. The interval lengths are then considered for all the tokens of the words $w \in W_\psi$.

The bottom half of Fig. 7.1 illustrates this scheme. Previously, only the word "Romeo" was considered in the analysis of interval lengths. Instead, the analysis in the lower part of Fig. 7.1 considers two words {"Romeo," "wherefore"}—a set of words—at the same time. In this case, the intervals are 1, 1, and 3 (instead of 1 and 4 for just "Romeo"), because "wherefore" occurs right after the second "Romeo," and the third "Romeo" occurs as the third word after "wherefore." As rare words occur only in small numbers, grouping multiple rare words together as a set serves to quantify their behaviors as an accumulated tendency.

Given a sequence of length m, the number of intervals is taken for one-ψth of the total number of words. Let W_ψ be the set of words in this sequence. For the above example, $m = 7$, $\psi = 1.75$ (as 4 out of 7 words were considered), and $W_\psi = \{$"Romeo," "wherefore"$\}$. Here, ψ is a real number because m is small, but in actual cases, we can specify that ψ is an integer. Hence, the lengths of all intervals of words in W_ψ can be analyzed. W_ψ then has the following interval sequence:

Fig. 7.5 Rare words considered as a set in *Moby Dick*

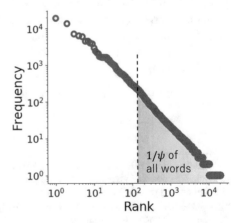

$$Q_\psi = Q_1, Q_2, Q_3 \dots, Q_i, \dots, Q_{\frac{m}{\psi}-1}, \qquad (7.7)$$

where Q_i is the interval in the original X between the ith and $(i+1)$-th occurrences of two (identical or different) elements of W_ψ, for $1 \le i \le \frac{m}{\psi} - 1$. The length of the interval sequence for all words of Q_ψ is $\frac{m}{\psi} - 1$, because there are m/ψ words in total. Similarly to the case of individual words, $P(q)$ and $S(q)$ are acquired from Q_ψ. The rest of this chapter considers this distribution of $Q_i = q$ in Q_ψ, while Chap. 8 deals with the interval sequence Q_ψ itself.

The technique of considering rare events to be of a single type is often adopted in extreme-value analysis of rare events such as large earthquakes or heat waves. This sort of analysis treats the intervals between rare events among a set of similar but less rare events, such as large earthquakes among the set of all earthquakes, by setting a certain threshold (corresponding here to ψ). The correspondence with interval analysis has also been reported for analysis of numerically extreme events, such as devastating earthquakes, and in the natural sciences and finance (Corral, 2004, 2005; Bunde et al., 2005; Santhanam and Kantz, 2005; Blender et al., 2014; Turcotte, 1997; Yamasaki et al., 2005; Bogachev et al., 2007). Here, we have assumed that the rare words in a language sequence correspond to extreme events. Moreover, although it does not make this correspondence with extreme-value analysis, the field of computational linguistics has used a similar idea to handle words occurring below a certain frequency as a single word.[7] Besides borrowing this technique, we will not delve any further into extreme-value analysis.

7.6 Behavior of Rare Words

The behavior of rare words can now be analyzed as in Sect. 7.4. Figure 7.6 shows the analysis for $\psi = 16$ words[8] in *Moby Dick* (left) and a shuffled text (right).[9] The exceedance probability for the shuffled text follows the black line, as the return intervals of rare words in that case behave in an exponential manner, similarly to "the." On the other hand, the plot for *Moby Dick* presents a clear deviation from the exponential function.[10] The value of θ in this case is 0.82. The results thus confirm the different behaviors of rare words in a real text and a shuffled text.

---

[7]Chapter 17 introduces that idea, but in brief, the idea is to consider the rare words of a corpus as a single word, as suggested in Mikolov et al. (2010).

[8]Tanaka-Ishii and Bunde (2016) investigated the effects of different values: $\psi = 1, 2, 4, 8, 16, 32, 64$. For a large ψ, the interval sequence becomes short, whereas for a small ψ, W_ψ starts to include functional words. Therefore, this book uses $\psi = 16$ as a moderate value throughout.

[9]$\varepsilon = 0.00787$ for *Moby Dick*, and $\varepsilon = 0.00521$ for the shuffled text.

[10]Tanaka-Ishii and Bunde (2016) indicated that κ is in fact a function of θ, and therefore, the stretched exponential function can ultimately be described by one parameter. Moreover, by using $P(q)$, they formulated a probability function for the occurrences of rare words, Q_ψ.

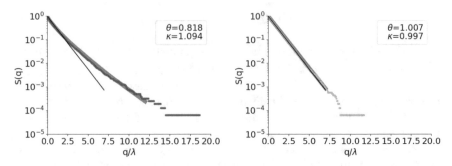

Fig. 7.6 Exceedance probability $S(q)$ for the rare words amounting to one-ψth of the text length m, with $\psi = 16$, for *Moby Dick* and a shuffled text. The black lines have a slope of -1, and the thick gray lines represent the fitted stretched exponential function

Tanaka-Ishii and Bunde (2016) used this approach to analyze 10 of the literary texts described in Chap. 5 in terms of ψ. They found that $S(q)$ showed a stretched exponential tendency for the rare words in all of the texts, and they concluded it to be a statistical universal across texts. Moreover, by applying this analysis to all of the literary texts mentioned in Chap. 5, we can find the same stretched tendency for all of them. This evidence further suggests that the bias in the interval length distribution is a statistical universal characterizing language.

Given that this is an accumulated behavior for rare words, we might suppose that the occurrences of any word in a text may behave similarly over a long term. The stretching factor indicates that in a real text, there is a chance that a word will return again even though the interval could be very large. Very rare words, even ones that do not occur even once in a text, nevertheless, do return in a set of texts.

One reason why rare words return in a text lies, of course, in human memory, as people do not typically forget many words that they have used. For example, even today we still use ancient Greek words, such as *hapax legomena* in this book. With an exponential function, the return probability should be almost zero, but in natural language texts, rare words return with a certain probability. Note that the *hapax legomena* (see Chap. 4) in one text often occur in other texts, as well. A set of texts exhibiting the same hapax legomena would constitute an archive. The differences between the exponential and stretched exponential functions revealed in this chapter are evidence that language is not produced randomly without memory but rather may derive from human memory.

The stretched exponential function fits the exceedance probability for not only certain words but also the set of all the words in a text. Figure 7.7 shows $S(q)$ for all word intervals in *Moby Dick* and its shuffled and monkey texts. These graphs thus show the exceedance probabilities for the graphs in Fig. 7.2.[11] For *Moby Dick*, there is a slight difference between the stretched exponential function (the gray fitted

---

[11] $\varepsilon = 0.0245, 0.0199, 0.0260$ for *Moby Dick*, the shuffled text, and the monkey text, respectively.

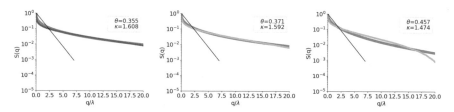

Fig. 7.7 Exceedance probability $S(q)$ for all interval lengths in *Moby Dick* (left), a shuffled text (middle), and a monkey text (right). The black lines have a slope of -1, and the thick gray lines represent the fitted stretched exponential function

line) and the real plot, but the overall fit is fairly good. On the other hand, the fitted functional parameters are $\theta = 0.36$ and $\kappa = 1.6$; thus, the exponent for the case of all words in *Moby Dick* is rather different from the exponents for "the" and the set of rare words. The stretching effect is deemed to also result from the aggregation of all words populated according to a power law.

As for the random texts, the plot for the shuffled text is very similar to the one for the real text, and it fits fairly well to the stretched exponential, with $\theta = 0.37$ and $\kappa = 1.6$. As mentioned above, however, the shuffled text remains hard to distinguish from *Moby Dick*. This similarity reflects the constraint that the total number of word occurrences in the shuffled text equals that of *Moby Dick*. On the other hand, in the right graph, the plot of $S(q)$ for the monkey text remains similar to the plot for *Moby Dick* only until a certain q. The monkey-text plot begins to deviate from the real plot for larger values of q, and it does not fit the stretched exponential.

$P(q)$ takes the following form when $S(q)$ is a stretched exponential:

$$P(q) \propto q^{\theta-1} \exp(-\kappa(q/\lambda)^{\theta}), \tag{7.8}$$

which is the probability density function of the Weibull distribution. It is similar to the function in formula (6.3), except that the stretching factor θ appears in both the exponential and power terms.

Section 6.1 and this chapter have explained the relations among the different views on population, i.e., the density function, the rank-frequency distribution, and the interval distribution. These relations further suggest that formula (7.8) could be used as a model of the population, and some previous works used the formula to model the rank-frequency distribution. For example, Nabeshima and Gunji (2004) used it to model the rank-frequency distribution of Japanese characters. For the large-scale data discussed in Sect. 5.3, however, Gerlach and Altmann (2013) reported that a double power law fits better than the Weibull function does. The Weibull function does not fit the Thomas data in CHILDES, either.

The analysis of return intervals in this chapter has again highlighted the population of elements in addition to the nature of memory in a text. We have observed the statistical properties of the population from different perspectives since Part II and found that the overall picture is only visible through pieces of evidence that highlight partial aspects of these properties. A unified mathematical view that incorporates all of these understandings remains an important future work.

Chapter 8
Long-Range Correlation

The previous chapter examined the return distributions of the words in a text. Another way to examine returns is in terms of how they succeed one another. As we will see here and in the following chapter, in a natural language text, a short return is likely to follow a series of short returns, and a long return is likely to follow a series of long returns. This causes a clustering phenomenon, meaning that at certain times, a word appears densely in a chunk of text, whereas at other times, the word hardly occurs. One source of such clustering phenomena in language lies in the context.

Various complex systems are known to exhibit similar clustering phenomena, as reported in the natural sciences and finance.[1] As natural examples, rainfall and earthquakes are known to occur in clusters. As for a social example, stock trades occur in clusters, as they trigger other trades. Likewise, a word triggers a set of other words, including itself, as this chapter will examine.

Methods for quantifying such clustering phenomena underlying a word sequence fall into two categories. The first kind of method examines the variance of the number of words that occurs within a range. This method seems promising because the presence of clusters suggests a large variance. Chapter 9 describes that approach, which we will refer to in a general sense as fluctuation analysis. The other kind of method, which is treated in this chapter, calculates the correlation between subsequences of a given sequence. It is based on the idea of measuring the degree of change in a sequence. The method is called long-range correlation analysis. Although the two approaches have certain similarities, they quantify different aspects of sequences, as summarized at the end of the next section. Here, we will see how the return sequence plays an important role in quantifying the degree of clustering in a sequence.

The original version of this chapter was revised. The correction to this chapter is available at https://doi.org/10.1007/978-3-030-59377-3_23

[1] Some examples were reported in Eisler et al. (2008); Corral (2004, 2005); Bunde et al. (2005); Santhanam and Kantz (2005); Blender et al. (2014); Turcotte (1997); Yamasaki et al. (2005); Bogachev et al. (2007).

© The Author(s) 2021, corrected publication 2022
K. Tanaka-Ishii, *Statistical Universals of Language*, Mathematics in Mind,
https://doi.org/10.1007/978-3-030-59377-3_8

8.1   Long-Range Correlation Analysis

Long-range correlation analysis considers how the correlation between two subsequences of a given sequence changes with respect to s, the distance between them, where $s \geq 0$. The analysis is based on the correlation function $C(s)$, which measures the correlation between the two subsequences. Figure 8.1 schematically illustrates this notion. Given a sequence, consider two subsequences at a distance s, indicated by the dashed and solid boxes in the figure; then $C(s)$ is the correlation between them. $C(s)$ can be computed by placing the dashed box at the start of the sequence and shifting the solid box toward the tail to measure $C(s)$ for larger values of s. We will formally define $C(s)$ later. Note that there are various correlation functions; typical ones are the mutual information and the autocorrelation function.

Our interest lies in how $C(s)$ changes with respect to s. When $s = 0$, $C(s)$ is the correlation between two identical sequences and should thus take the maximum value. When $s > 0$, $C(s)$ usually decreases as s increases. Some sequences typically have local similarities and produce correlation only for small s. Such a sequence is *short-range correlated*. When a sequence has some correlation that persists even at a long distance, however, $C(s)$ remains large even for a large s. Such a sequence is *long-range correlated*. Precisely, a sequence is long-range correlated when a correlation function $C(s)$ follows a power function with respect to the distance s between two of its subsequences:

$$C(s) \propto s^{-\gamma}, \quad s > 0, \quad 0 < \gamma < 1, \tag{8.1}$$

where $-\gamma$ is the power exponent indicating the degree of decay of $C(s)$ with respect to s. If $C(s)$ can be characterized as a power function, it implies that the correlation slowly decays with the distance s and thus remains substantial over long distances. Overall, the existence of long-range correlation roughly suggests that any two subsequences of a sequence have some similarity.

Not all correlation functions can capture the long-range properties underlying language, however, even when such phenomena exist. As we will see in the rest

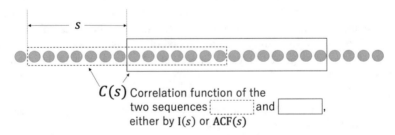

Fig. 8.1 Schematic illustration of long-range correlation analysis. The chain of gray circles represents a sequence (of words). The analysis computes the correlation function $C(s)$ between two subsequences at a distance s apart

of Part III, only the right choice of function can reveal the fluctuation underlying language, which appears to be a weak phenomenon.

8.2 Mutual Information

Because a natural language text forms a symbolic sequence, a natural choice of correlation function is the mutual information, which is defined as

$$I(s) \equiv \sum_{a,b} P(X_i = a, X_{i+s} = b) \log \frac{P(X_i = a, X_{i+s} = b)}{P(X_i = a)P(X_{i+s} = b)}, \qquad (8.2)$$

where X_i and X_{i+s} are elements separated by a distance s, and a and b indicate words or characters. As this function can be directly applied to a nonnumerical sequence, it is a good candidate to use as $C(s)$ for a text. Because the term in the logarithm gives the mutual information of a and b at a distance s, averaging over all possible values of a and b gives the average degree of co-occurrence of two elements at a distance s.

This function has been used to analyze DNA sequences (Li et al., 1994; Voss, 1992). Moreover, Li (1989), Ebeling and Pöschel (1993), and Lin and Tegmark (2017) have attempted to use it to study language, but there have been questions about its efficacy in quantifying the degree of long memory. Li (1989) indicated that the mutual information gave a power-like decay for letter sequences, but even then γ was very large. Accordingly, he stated that "*it is not conclusive as to whether the same inverse power law function extends beyond short distances.*" Lin and Tegmark (2017) showed a power-law decay for a natural language character sequence, but the data included chunks of other sequences besides natural language characters (such as a long sequence of numbers); Takahashi and Tanaka-Ishii (2017) concluded that their results cannot be generalized to a power-law decay for natural language.

Figure 8.2 shows the mutual information for *Moby Dick*, with the left graph showing the result for words and the right graph showing that for characters on double-logarithmic axes. The graph for words exhibits a very fast decay already at $s = 5$, indicating no long-range correlation. The plateau for $s > 10$ indicates that the distributions of word pairs are almost the same. The correlation for characters persists a little longer than it does for words, but the decay is still rapid. Many other natural language texts exhibit similar tendencies of the mutual information, for both characters and words. The mutual information does not exhibit a power-law decay for natural language because the number of types of elements, v, is too large, even in the case of characters, for the mutual information function to persist. On the other hand, the mutual information works well on data that has few types of elements, such as DNA sequences, in which v is only four, and music as processed by Lin and Tegmark (2017), which also has fewer types than language does.

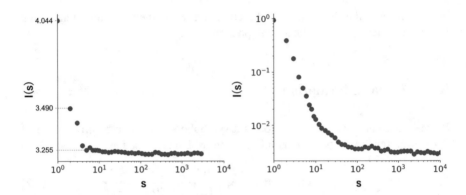

Fig. 8.2 Long-range correlation using the mutual information for *Moby Dick* in terms of words (left) and characters (right). The horizontal axis indicates the distance *s* between occurrences of the same word or character, while the vertical axis indicates the mutual information

Still, Lin and Tegmark (2017) proved analytically that a sequence produced by using a grammatical framework has mutual information that presents a power-law decay. This is an important fact from a language perspective, as grammatical structure is an important characteristic of natural language. We will return to this topic later, in Chap. 14.

8.3   Autocorrelation Function

The most commonly used correlation function for long-range correlation analysis is the autocorrelation function (ACF). The ACF is typically applied to numerical sequences, however, and it is not obvious how to apply it to natural language sequences, which are nonnumerical. Accordingly, we will have to determine how to transform a natural language text into a numerical sequence. First, suppose we have a numerical sequence $x_1, x_2, \ldots, x_i, \ldots, x_m$, where $x_i \in \mathbb{X}$, a numerical set, for $1 \leq i \leq m$. Let the sequence's mean and standard deviation be μ and σ, respectively. The autocorrelation function is defined as follows:

$$\text{ACF}(s) \equiv \frac{1}{(m-s)\sigma^2} \sum_{i=1}^{m-s} (x_i - \mu)(x_{i+s} - \mu). \tag{8.3}$$

This is a fundamental function to measure the correlation between two subsequences whose heads are separated by a distance *s*. In mathematical terms, ACF(*s*) calculates the covariance between the original sequence and a subsequence, normalized by the original variance σ^2. The value of ACF(*s*) ranges between -1 and 1, with ACF(0) = 1 by definition. For a simple random sequence, such as a random binary sequence, the function gives small values fluctuating around zero for any $s > 0$, as such a sequence has no correlation with itself. A sequence is judged as long-range correlated when ACF(s) decays by a power law, as specified in formula (8.1).

As the autocorrelation function is generally applicable only to numerical time series, it has been unclear how to apply it to a natural language text until recently. The main reason was the difficulty of appropriately transforming a natural language text into a numerical time series. Besides the ACF, various other functions that are applied to numerical sequences have been used with natural language sequences via arbitrary transformations into numerical sequences.

Here, we will briefly summarize these transformation approaches that enable analysis of natural language sequences. Montemurro and Pury (2002) transformed a word sequence into a sequence of ranks in order of the word frequency, while Kosmidis et al. (2006) transformed a word sequence into a corresponding sequence of the word lengths, but their transformations were essentially arbitrary. For instance, the difference in rank between the words ranked 1 and 2 does not necessarily have the same significance as that between the words ranked 1001 and 1002. Ebeling and Neiman (1995) transformed the letters in the *Bible, Grimms' Fairy Tales*, and *Moby Dick* into binary sequences of appearance/nonappearance, and Altmann et al. (2012) adopted the same approach for characters and character sequences. Although the ACF could be applied to those binary sequences, the resulting analysis would be similar to another approach, which is used here. Specifically, Tanaka-Ishii and Bunde (2016) proposed using return-interval sequences, which were introduced in Sect. 7.5. Because these sequences are numerical, they provide a natural way to transform a nonnumerical language sequence into a numerical sequence.

8.4 Correlation of Word Intervals

For the return intervals defined in Sect. 7.1, we first analyzed the return interval sequence for a particular word having a long sequence. That analysis method was only applicable to very frequent words. Therefore, as described in Sect. 7.5, we analyzed the interval sequence of rare words by grouping them in a set. Here, using the example of *Moby Dick*, we will follow the same approach by applying the ACF first to the return interval sequence of "the" and then to the return interval sequence of a set of rare words.

First, a hint can be gained from a previous study on how to correctly apply the ACF function to real sequences. Lennartz and Bunde (2009) showed that, in general, the autocorrelation function is only valid for values of s up to $|Q|/100$, where $|Q|$ is the length of the interval sequence. For s larger than $|Q|/100$, the sequence length becomes shorter relative to s, causing the autocorrelation function values to decrease rapidly. Moreover, these values often fluctuate at large s, making the decreasing

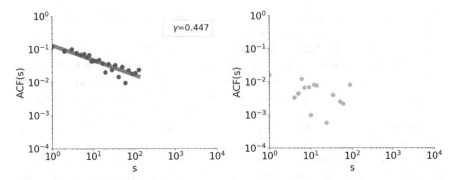

Fig. 8.3 Autocorrelation function of the returns of "the" in *Moby Dick* (left) and a shuffled text (right), plotted on double-logarithmic axes. The x-axis indicates the distance s between the number of occurrences of "the," and the y-axis indicates the value of the autocorrelation function, which was applied to the interval sequence of "the." In the left graph, the thick gray line is the fitted power function (formula (8.1)), and the value of γ is shown in the upper right corner. The range of correlation extends up to almost $s = 100$, but the *interval* sequence of "the" has a mean interval of 18.5, so the occurrences of "the" are actually correlated over a distance of at least $18.5 \times 100 = 1850$ words

tendency hard to see. Therefore, we will limit the analysis to the range of $s \leq |Q|/100$; in addition, we will use a logarithmic bin.[2]

As a result, Fig. 8.3 shows the autocorrelation function for the word "the" on double-logarithmic axes, for *Moby Dick* on the left and a shuffled text on the right. In the left graph, the plot follows a straight line,[3] with $\gamma = 0.45$. All the ACF(s) values are positive. The plot shows that the autocorrelation of the interval sequence has a value above 0.1 at $s = 1$ and then decays slowly; it is also nearly straight, thus showing a power-law decay. This behavior continues until $s \approx 100$, and the correlation remains well above 0.0. Because "the" occurs 13,772 times, the mean interval length is 18.5 words. Accordingly, the strong correlation at $s = 100$ means that the behavior of "the" is correlated even for sequences 1850 words apart. In other words, the influence of how "the" occurs is persistent even at a distance of a few thousand words. Nevertheless, γ for the interval sequence is almost 0.45; this rather large value implies a tendency for the interval lengths of "the" to become dissimilar with increasing s, despite the plot presenting a power-law decay. This is the case because "the" is a functional word.

On the other hand, the graph for the shuffled text looks very different from the one for *Moby Dick*. Because ACF(s) values range between -1 and 1, points with

---

[2] A logarithmic bin is a range that extends exponentially along s. This is a common technique for analysis of power function decay (Clauset et al., 2009). Although Part II did not apply it, a logarithmic bin is commonly used whenever some points appear in a cloud, as shown in the figures in Sects. 6.1 and 7.1. For analysis with the autocorrelation function, this book uses an integer bin in the range of $1 \leq s \leq 10$ and after $s = \lceil 10 \times 1.2^k \rceil$, $k = 1, \ldots$.

[3] For fitting Fig. 6.3, the least-squares method was applied (cf. Sect. 21.1). $\varepsilon = 0.00884$.

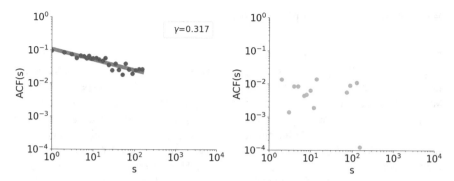

Fig. 8.4 Autocorrelation function ACF(*s*) on double-logarithmic axes for the rare words of *Moby Dick* (left) and a shuffled text (right), for an interval sequence of rare words consisting of 1/16th of all the words. In the left graph, the long-range correlation persists up to $s = 100$, which means that the correlation in fact persists to a distance of 100 times the mean interval length, or 1600 words apart

ACF(*s*) ≤ 0 cannot be shown on double-logarithmic axes. As a result, this graph shows a smaller number of points, because many are in fact negative and thus not shown. A point with a negative value for ACF(*s*) indicates an anticorrelation. Here, however, the negative values are close to zero. When both positive and negative values are small, it indicates that there is *no* correlation. In the figure, indeed, the positive points with ACF(*s*) > 0 in the right graph (for the shuffled text) have smaller values than those in the left graph (for *Moby Dick*).

The sequence of intervals between occurrences of "the" is therefore long-range correlated, yet the decay in the correlation is rather fast. The question now is what will we find for rare words. As mentioned above, we can apply the autocorrelation function to a sequence of intervals between rare words by grouping multiple rare words in a set. That is, we consider multiple rare words forming an interval sequence Q_ψ. Similarly to Fig. 8.3, Fig. 8.4 shows the results on double-logarithmic axes for the case of $\psi = 16$. The red points in the left graph represent the ACF(*s*) values for *Moby Dick* in its entirety. All the ACF(*s*) values here are positive, and they show power-law behavior.

The value of γ in this case is 0.32;[4] thus, the decay is slower than in the case of "the," meaning that the correlation persists longer. Moreover, the plot is straight until $s = 100$. Here, the interval sequence consists of m/ψ elements of the original sequence X, and therefore, the mean interval length is $m/(m/\psi) = \psi$. As $\psi = 16$, this indicates that the rare words in this natural language text are correlated at least up to a distance of 1600 words apart.

On the other hand, the right graph of Fig. 8.4 shows that the rare words in the shuffled text are not correlated. In fact, ACF(*s*) is small and negative for many values of *s*. Note that a monkey text would not show any correlation for rare words either.

---

[4] $\varepsilon = 0.00761$.

8.5   Nonstationarity of Language

Now let us examine the implication of these empirical results. Given an interval sequence $Q = Q_1, Q_2, \ldots, Q_i, \ldots, Q_{|Q|}$, with $1 \leq i \leq |Q|$, let us assume for now that the mean and variance of Q_i exist; we denote them, respectively, as μ and σ^2. Let $E[Q_i]$ denote the expected value of Q_i, and let $\text{Cov}[Q_i, Q_{i+s}]$ denote the covariance of two subsequences starting at Q_i, Q_{i+s}.[5] The covariance is nearly equivalent to the autocorrelation function seen thus far, differing only by a constant factor; specifically, the autocorrelation thus far is equal to $\text{Cov}[Q_1, Q_{1+s}]/\sigma^2$. In other words, we always fixed i at 1 and measured the correlation by varying s. This scheme can be generalized so that i can be moved, too. In the context of Fig. 8.1, we can place the dashed box at the starting element of Q_i and the solid box at distance s apart. Then, σ^2 times the autocorrelation value gives the covariance $\text{Cov}[Q_i, Q_{i+s}]$.

A stochastic numerical sequence Q is said to be *stationary* when $E[Q_i]$ and $\text{Cov}[Q_i, Q_{i+s}]$ are *constant* across the sequence (Pipiras and Taqqu, 2017; Shumway and Stoffer, 2011). Precisely, for any i,

$$E[Q_i] = \mu, \tag{8.5}$$

$$\text{Cov}[Q_i, Q_{i+s}] = \sigma^2. \tag{8.6}$$

The signification of this definition is that a sequence with stationarity has *non-changing behavior* everywhere in the sequence. This concept of stationarity can be used as a criterion to examine the degree of *changing* behavior for a language sequence, as seen thus far.

Previously, we analyzed ACF(s) by changing s. To verify the stationarity from the covariance point of view, the value of ACF(s) should be analyzed by varying i. Fixing s at some value and changing i produces fluctuation of the ACF(s) value locally and globally, for the interval sequences both of "the" and of rare words. This analysis provides some evidence that the interval sequences of natural language are not stationary.

The above definition of stationarity is considered too strict for some purposes; a more common notion these days is that of a sequence being *weakly stationary* (Pipiras and Taqqu, 2017; Shumway and Stoffer, 2011):

$$E[Q_i] = \mu, \tag{8.7}$$

$$\text{Cov}[Q_i, Q_{i+s}] = \text{Cov}[Q_1, Q_{1+s}]. \tag{8.8}$$

---

[5]$\text{Cov}[Q_i, Q_{i+s}]$ is defined as follows:

$$\text{Cov}[Q_i, Q_{i+s}] \equiv E[(Q_i - \mu)(Q_{i+s} - \mu)] \tag{8.4}$$

Here, the second line means that any decay in the covariance of subsequences is the *same* everywhere. As mentioned above, until the previous section, we only examined the case of $i = 1$. In the case of $i > 1$, we would find that the covariance empirically presents a power-law decay for natural language, as well. Nevertheless, it is difficult to verify whether γ varies with i. The fact that the autocorrelation function is valid only up to $s < |Q|/100$ constrains the range of i even further.

Given that the covariance of a language sequence presents a power-law decay everywhere, it might seem reasonable to say that the interval sequences of "the" and rare words possess some quality of being weakly stationary. However, natural language very likely deviates from being even weakly stationary, for at least the two following reasons.

First, the analysis in the previous section treated a set of rare words as a single word; in reality, rare words are different. This fact already makes the sequence nonstationary. Furthermore, the mean and variance values exist for Q_ψ, but for an interval sequence of a rare word, we saw in the previous section that its interval length could be arbitrarily long. Accordingly, it is not guaranteed that the mean and variance are defined.

Second, natural language does not fulfill the condition given in formula (8.7) for the mean behavior of $E[Q_i]$. For the shuffled text considered above, $\mu = 18.5$, and for rare words the mean is $\mu = \psi = 16$ and uniform across the text. On the other hand, for *Moby Dick*, the mean value for "the" or even for rare words fluctuates. Even the occurrences of "the" are clustered: they are abundant in some parts of the text but rare in other parts. This suggests a complex pattern underlying occurrences of "the." Therefore, $E[Q_i]$ cannot equal μ for all i.

Figure 8.5 shows the occurrences of "the" within the first 5000 words of *Moby Dick*. Each vertical bar indicates a location where it occurs. Note that it is difficult to see a regular pattern in this sequence. The vacant regions, indicating no occurrences of "the," are separated by regions where "the" occurs densely, in clusters. These recurring waves of clusters of "the" in the text cause the correlation to persist. One source of clustering phenomena of "the" is changes of context. Within an English paragraph, we introduce general ideas with indefinite articles. When we elaborate those ideas, we use definite articles. Context changes should also emphasize the clustering phenomena of rare words.

Tanaka-Ishii and Bunde (2016) presented another analysis method to quantify clustering phenomena. They showed that when there are k short intervals, the $(k + 1)$-th interval is likely to be short. On the other hand, a long interval is likely to follow k successive long intervals. Therefore, words appear clustered in a natural language text.

Consequently, the occurrences of words possess a complex pattern, intuitively verbalized thus far by the term *clustering* behavior. One conclusion drawn in

Fig. 8.5 Occurrences of "the" in the first 5000 words of *Moby Dick*

this chapter is that natural language is nonstationary, which is highlighted by the behavior of rare words and the clustering phenomena of words in general. We will need future efforts to gain a mathematical understanding of how precisely nonstationary natural language is.

8.6   Weak Long-Range Correlation

Tanaka-Ishii and Bunde (2016) analyzed γ for 10 literary texts across languages and showed that the γ values were similar for different values of ψ. Moreover, the same tendency was found (Tanaka-Ishii, 2018) for infant utterances and music, as described in Chap. 5. Then, how universally does long-range correlation hold across texts?

It is difficult to judge whether a text is long-range correlated by using the method presented in this chapter. One potential approach is to examine the number of negative values for s. When there is no correlation at a given s, the autocorrelation function values become very small, with half being negative. In the case of *Moby Dick*, no negative values occur within the scope shown in Figs. 8.3 and 8.4.

This is not always the case, however: some literary texts do present at least one small negative autocorrelation value for some s values. In testing the 1142 literary texts analyzed in Chap. 5, 201 texts had a few small negative autocorrelation values. The majority of them occurred at $s > 50$, but one text had small negative values at $s < 10$, and 27 texts had them at $s < 20$. Therefore, 17.6% of the texts showed some weak long-range correlation. Even for texts that did not present any small negative values, the strength of ACF(s) varied, often with a large γ. We can thus say that the long-range correlation underlying a text is universal only to the extent outlined by these cases.

Nevertheless, we cannot immediately conclude that 17.6% of those literary texts are not long-range correlated. The proportion of small negative points was far less than half, and it seems that the long-range correlation could be weak, and that the function ACF(s) has a limitation in measuring such weak long memory in a text. This chapter has shown that the mutual information function could not quantify the long-range correlation of natural language texts, but this does not mean that long-range correlation does not exist in natural language. Rather, the mutual information function is not capable of detecting it even when it exists. The mutual information is effective for DNA sequences, but not for natural language. Similarly, although the autocorrelation function can detect weak long-range correlation, it might not be able to do so on some texts. Our analysis thus far suggests that the long-range correlation phenomena of natural language should be deemed weak, and their quantification requires us to choose the right correlation function $C(s)$.

Apart from this issue, our discussion also emphasizes the problem of methodology. The essential problem is that this method of long-range correlation analysis cannot clearly judge whether phenomena exist. That is, although we have shown that the autocorrelation function is more sensitive than the mutual information function, a contrarian critique would state that phenomena that might not actually exist. The difference between the real and shuffled texts compared in this chapter could be a matter of degree.

The overall consequences also show the difficulty of applying the numerical autocorrelation function to a nonnumerical sequence. Quantifying clustering behavior requires a different perspective to complement this chapter's observations. As it turns out, the next chapter applies another method that supports the overall conclusion in this chapter: that the long memory derived from clustering behavior is *weak*.

Chapter 9
Fluctuation

The previous two chapters presented analyses based on return intervals. Chapter 7 was about the distribution of returns, whereas Chap. 8 considered sequences of returns.

This chapter sets aside returns and introduces another genre of methods, called fluctuation analysis. These methods can also be used to investigate the clustering phenomena underlying word sequences. If a word occurs in a clustered manner, then its counts within a certain window should vary greatly. This idea leads to two main analysis methods, which both yield power functions.

9.1 Fluctuation Analysis

When events occur in a clustered manner, some periods have no events while other periods contain many events; i.e., the events occur densely in these periods. Figure 8.5, for example, showed periods devoid of "the," as well as periods in which "the" occurred densely. This feature suggests that we should examine the variance of the number of occurrences in an interval of text. We refer to the use of methods that examine the variance of events as *fluctuation analysis*.

We can trace the origin of fluctuation analysis back to a method developed by Hurst (1951). Hurst was interested in predicting the flow of the Nile River in relation to environmental occurrences such as rainfall. Rainfall occurs in a clustered manner, often causing floods, and prediction of such events is thus an important issue. With respect to a time span l, Hurst proposed to observe the difference between the maximum and minimum water levels, denoted as $d(l)$ (normalized by the deviation of the inflow amount). He found that $d(l)$ grows with respect to l according to the following power function with an exponent χ:

$$d(l) \propto l^{\chi}. \tag{9.1}$$

© The Author(s) 2021
K. Tanaka-Ishii, *Statistical Universals of Language*, Mathematics in Mind,
https://doi.org/10.1007/978-3-030-59377-3_9

The function $d(l)$ measures a kind of variance of the water level. The analysis implies that the variation in water level increases according to a power function of the time span. Hurst's result implies that flooding or drought of any arbitrarily devastating degree can occur given enough time.

Montemurro (2001) applied Hurst's method to the clustering phenomena underlying language. Given that the method is applicable only to numerical time series, they had to transform a natural language text into a numerical time series, like we did in the previous chapter on the autocorrelation function. As mentioned in Sect. 8.3, they transformed a word sequence into a sequence of ranks ordered by the word frequency. They then measured the difference between the maximum and minimum ranks and showed how the (normalized) difference in rank follows a power law with respect to the time span. Although there was an arbitrariness in their transformation of a word sequence into a sequence of ranks, it was an interesting idea to compare a linguistic word sequence to the Nile's flow.

It might be interesting to apply Hurst's method to an interval sequence, similarly to our autocorrelation approach. However, Hurst's method has limitations, including the fact that it does not use the statistical variance but rather only the difference between the maximum and minimum values in an interval, which weighs outliers too heavily. Moreover, language essentially consists of a nonnumerical sequence of linguistic elements, and methods analyzing the variance can be applied directly, even without an arbitrary transformation into a numerical sequence. Hence, the rest of this chapter considers analysis methods that apply to nonnumerical sequences.

Ebeling and Neiman (1995) analyzed the power-law relation between the lengths of segments and the variance of the number of characters in the segments. Theirs was one of the first applications of fluctuation analysis to natural language. Their method constitutes a natural, intuitive approach. In the following, we will call their method the EN method or EN analysis.

Given a set of elements (characters in the original work), denoted as C, let $y(c, l)$ be the count of element $c \in C$ within segments of length l, with $1 < l < m$, where m is the sequence length, and let $\text{Var}[y(c, l)]$ be the variance of the count $y(c, l)$ across segments of length l. Then, the fluctuation function $y(l)$, which has been empirically shown to grow by a power function with respect to l, is defined as follows:

$$y(l) \equiv \sum_{c \in C} \text{Var}[y(c, l)] \propto l^{\nu}, \qquad \nu > 0. \qquad (9.2)$$

This book calls ν the EN exponent.

Theoretically, $\nu = 1$ for an i.i.d. sequence.[1] Ebeling and Neiman (1995) reported that the characters of the *Bible* have $\nu = 1.69$, indicating significant fluctuations in the occurrences of characters.

---

[1]The number of times an element appears in a segment l in an i.i.d. sequence follows a Poisson distribution, as deduced in Sect. 21.5. A Poisson distribution is characterized by its mean and variance being equal. If l is made k times larger, then its mean obviously becomes k times larger as well, and so does the variance. Therefore, $\nu = 1$ for an i.i.d. sequence. Section 21.6 gives a more formal explanation.

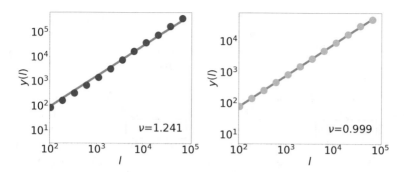

Fig. 9.1 Ebeling and Neiman's fluctuation analysis applied to *Moby Dick* (left) and a (word-) shuffled text (right). Both graphs show results for *character* sequences

The left graph of Fig. 9.1 plots $y(l)$ for the character sequence of *Moby Dick*. The right graph shows the result for a (word-) shuffled text, which has $\nu \approx 1$.[2] Although it is not shown here, the result for a monkey text is very similar to the one for the shuffled text. Given that a shuffled text has average fluctuations, and that the fluctuations in a real text are larger than in the random case, we would expect ν to be larger than 1 for *Moby Dick*; indeed, as shown in the left graph, $\nu = 1.24$.

Although Ebeling and Neiman (1995) did not report such an application, their method is applicable to word sequences, too. In this case, the variance over the set of characters $c \in C$ is replaced by the variance over words $w \in W$ in formula (9.2). For the word sequence of *Moby Dick*, $\nu = 1.25$, almost the same as for characters.[3] Note that the similarity of the ν values for characters and words is probably only a coincidence.

Apart from language, fluctuation analysis methods analyzing the variance with respect to l have been applied to nonnumerical time series in various other fields, though not as frequently as to numerical time series. The analysis often suffers from noise that depends locally on previous values. For this reason, time series are often "detrended" by a linear function to remove overall trends in their values (Peng et al., 1994; Kantelhardt et al., 2002).

9.2 Taylor Analysis

Another way to quantify fluctuations is via Taylor analysis (Taylor, 1961). Taylor's law characterizes how the standard deviation of the number of events for a given

---

[2]For fitting Fig. 9.1, the least-squares method was applied (cf. Sect. 21.1) to the plots for $l \propto a^k$, with $k = 1, 2, 3, \ldots$, $a = 1.8$. $\varepsilon = 0.0797$ for *Moby Dick*, whereas $\varepsilon = 0.00936$ for the shuffled text. As introduced in Sect. 3.5, the shuffled text used here is word shuffled, and therefore, the order of the characters within words is maintained. Nevertheless, $\nu \approx 1$.

[3]$\varepsilon = 0.0541$.

time or space grows with respect to the mean. The EN method accumulates the variance with respect to l, despite events being power distributed, as this section will show. In contrast, Taylor analysis directly analyzes the relation between the standard deviation (or the variance, which produces equivalent results) and the mean of the events, prior to EN analysis.

The pioneering study of this concept was reported in 1938 (Smith, 1938), well before Hurst's method. H. Fairfield Smith investigated the fluctuations underlying crop yields. The idea was rediscovered by Taylor (1961), who measured the sizes of various biological colonies. One characteristic of Taylor's law is its capability to distinguish different categories of data (Taylor, 1961; Taylor, 2019). Taylor analysis has thus been used in many fields, including ecology, life science, physics, finance, and human dynamics, as well summarized by Eisler et al. (2008). However, it has hardly been used to study natural language, besides Gerlach and Altmann (2014), who used it to study the mean and variance of the *vocabulary size* of texts. A more basic way to apply Taylor analysis is based on event occurrences, by considering the standard deviation for every type. Hence, this section describes previous findings using that approach (Kobayashi and Tanaka-Ishii, 2018; Tanaka-Ishii and Kobayashi, 2018).

As mentioned above, Taylor's method considers how the standard deviation changes with respect to the *mean* of occurrences, given a time length. That is, for a given segment length l, the number of occurrences of a specific word $w_k \in W$ is counted for every segment, and the mean μ_k and standard deviation σ_k across segments are obtained. In the notation of the EN method, $\mu_k \equiv \mathrm{E}[y(w_k, l)]$, while $\sigma_k^2 \equiv \mathrm{Var}[y(w_k, l)]$. Figure 9.2 illustrates this procedure schematically. The chain of circles represents a sequence of words, with a particular word represented by the black circles. When the word's occurrences fluctuate, it appears abundantly in some boxes but sparsely in other boxes.

Calculating the mean μ and standard deviation σ for all word types $w_1, \ldots, w_k, \ldots, w_v \in W$ gives the distribution of σ with respect to μ, where v indicates the number of element types in the set W and is not to be confused with v in the preceding section. Following a previous work (Eisler et al., 2008), we will define Taylor's law as the relationship that holds when μ and σ are correlated by a power function with exponent α, as follows:

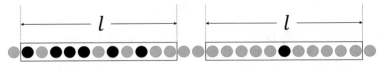

Compute the mean and standard deviation of ● in boxes of length l

Fig. 9.2 Schematic illustration of Taylor analysis. The chain of circles represents a sequence of words, with a particular word represented by the black circles. The analysis computes the mean and standard deviation of the frequency of the word within a distance l in the text. When the word fluctuates, it appears frequently in some boxes but hardly at all in other boxes

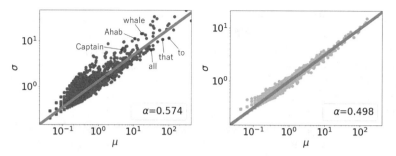

Fig. 9.3 Taylor analysis of *Moby Dick* (left) and a shuffled text (right). Each point represents a type of word. The values of the standard deviation σ and mean μ for each word type are plotted for a segment size of $l = 5620$. The Taylor exponent obtained for *Moby Dick* by the least-squares method is 0.57, whereas that for the shuffled text is 0.5, as theoretically expected for an i.i.d. process. The left graph highlights some keywords in *Moby Dick*

$$\sigma \propto \mu^{\alpha}, \qquad \alpha > 0. \tag{9.3}$$

Even for an i.i.d. process, this formula holds with $\alpha = 0.5$, as explained below. For real data, it is empirically known that $0.5 < \alpha < 1$, and we are especially interested in this difference in α from the i.i.d. case. In other words, Taylor's law suggests that an i.i.d. process naturally fluctuates as a statistical consequence, while real processes fluctuate more strongly than would be expected for a random process.

Figure 9.3 shows Taylor analysis graphs on double-logarithmic axes of the distributions of words in *Moby Dick* and a shuffled text. The horizontal axis indicates the mean frequency of a specific word within a segment of $l = 5620$ words, whereas the vertical axis indicates the standard deviation.[4] For *Moby Dick*, the points are distributed around the regression line, thus showing a power-law tendency. The exponent[5] is $\alpha = 0.57$. Although the global tendency follows a power law, many points deviate significantly from the regression line. The words with the greatest fluctuations are often keywords. For example, among words in *Moby Dick* with large μ values, those with the largest σ values include some of the book's most important keywords, such as "whale," "captain," and "Ahab," whereas those with the smallest σ values include functional words such as "to," "that," and "all."

In general, α is small when the fluctuations are small and large when the fluctuations are large. It has been analytically proven that $\alpha = 0.5$ for an i.i.d.

---

[4] The choice of l is arbitrary but must be sufficiently smaller than the text length to accurately calculate the mean and standard deviation. Among different values of l taken from logarithmic bins, a maximum l that could apply to all texts was adopted here. This resulted in a choice of l around 5000 words, specifically $l = 5620 \approx 10^{3.75}$.

[5] For fitting Fig. 9.3, the least-squares method was applied (cf. Sect. 21.1). $\varepsilon = 0.0623$ for *Moby Dick*, whereas $\varepsilon = 0.0238$ for the shuffled text.

process (Yule, 1944; Kobayashi and Tanaka-Ishii, 2018).[6] As shown in the right graph of Fig. 9.3, this is the case for the shuffled text. Although not shown here, the same is true for a monkey text. Moreover, when words of one type, which occur only a limited number of times, appear once in a segment of length l, $\alpha = 0.5$; however, when they all appear in one segment, $\alpha = 1$ (Tanaka-Ishii and Kobahashi, 2019). Therefore, bursty word occurrences tend to raise the value of α.

One case with $\alpha = 1$ occurs when all segments *always contain the same proportions of the elements of W*. For example, suppose that $W = \{a, b\}$. If b always occurs twice as often as a in all segments (e.g., three a and six b in one segment, two a and four b in another, etc.), then both the mean and standard deviation for b are twice those for a, so the exponent is 1. In a real text, this cannot occur for all W, so $\alpha < 1$. Nevertheless, for a subset of words in W, this could happen, especially for a regular grammatical sequence. For instance, consider a programming statement: while (i < 1000) do i++. Here, the words while and do always occur once in this type of statement, whereas i always occurs twice. This example shows that the exponent indicates how consistently words in W depend on each other, i.e., how words co-occur systematically in a coherent manner, thus suggesting that the Taylor exponent is partly related to grammaticality.

The Taylor exponent thus partly indicates the degree of consistent co-occurrences among words. Though well above 0.5, the value of 0.57 obtained here suggests that the words of natural language texts are *not so strongly* or consistently coherent with respect to each other. That is, the fluctuations in a language sequence are not strong, being closer to the i.i.d. case rather than presenting a strong clustering phenomenon. This understanding coincides with the conclusion of the previous chapter that the long-range correlation in language is weak.

Figure 9.4 summarizes the overall picture obtained by applying Taylor analysis to all the literary texts examined in Sect. 5.1 and to many other texts listed in Sect. 22.3. The horizontal axis indicates the kind of data, and the vertical axis indicates the value of α. Here, each point represents one data source, and the median and quantiles of the Taylor exponents are shown for each kind of data.

The first plot shows the result for a set of shuffled data; all exponents are very close to 0.50. The remaining plots for real data all have exponents higher than 0.5. Such a clear distinction between the shuffled and real texts is an advantage of this method over the long-range correlation approach, whose limitations were discussed at the end of Chap. 8.

In Fig. 9.4, the second and third plots from the left contain the exponents for the literary texts examined in Sect. 5.1, and the fourth plot contains the exponents

---

[6]Previously, the corresponding value for the EN method was 1. Taylor analysis uses the standard deviation, so the i.i.d. case has $\alpha = 0.5$. Section 21.6 gives a more formal explanation.

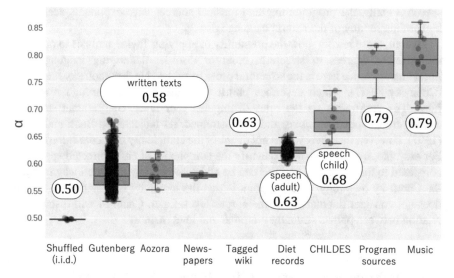

Fig. 9.4 Box plots of the Taylor exponents for different kinds of data. Each point represents one data sample, and samples from the same kind of data are contained in each box plot. The first plot is for shuffled data, while the remaining plots represent real data, including both natural language texts and language-related sequences. Each box ranges between the quantiles, with the middle line indicating the median, the whiskers showing the maximum and minimum, and some extreme values lying beyond. The values indicated above or below the plots are the mean exponents for each data kind

for the newspapers.[7] The average result is 0.58 for these written texts. The results for Project Gutenberg range from 0.53 to 0.68, but the number of texts starts to decrease significantly at a value of 0.63, showing that the distribution of the Taylor exponent is rather narrow. The kinds of texts at the upper limit of the exponent include structured texts of fixed style, such as lists of histories and *Bibles*. A detailed nonparametric test of the literary texts showed that different languages were not distinguishable by the value of α.

The last five plots in Fig. 9.4 are for tagged wiki data (a corpus of Wikipedia data), speech data (the Japanese Diet's spoken record; and CHILDES, containing child-directed speech), programming language source code, and music data. In these cases, the Taylor exponents clearly differ from those of the written natural language texts. The tagged wiki dataset shows a higher exponent of 0.63 because of the consistent format of the wiki tags. The speech data also shows higher exponents than those of the written texts, 0.63 for the Diet record and 0.68 for the child-directed

---

[7]The fourth plot shows the Taylor exponent for the *Wall Street Journal* in addition to the Taylor exponents for the Japanese and Chinese newspapers used in (Kobayashi and Tanaka-Ishii, 2018; Tanaka-Ishii and Kobayashi, 2018).

speech. Finally, the programming and musical sources have exponents near 0.79, higher than for any of the natural language data.

Overall, these results show the possibility of applying Taylor analysis to quantify the different degrees of structural order or disorder underlying language. An important reference here is the structural order formulated in the Chomsky hierarchy (Chomsky, 1957), which describes phrase structure grammar through rewriting rules. The constraints on the rules distinguish four levels of grammar: regular, context-free, context-sensitive, and unconstrained. As indicated in Badii and Politi (1997), however, that theory does not quantify the complexity on a *continuous* scale. For example, we might want to quantify the complexity of child-directed speech as compared to that of adults, and this can be addressed in only a limited way through the Chomsky hierarchy. Another point is that the hierarchy is sentence based and does not consider fluctuations at the corpus level. Later, Chap. 14 will discuss the relation between phrase structure grammar and long memory.

9.3 Differences Between the Two Fluctuation Analyses

The methods and results of EN and Taylor analyses tend to be similar because both are based on $\mathrm{Var}[y(w, l)]$. Nevertheless, the two methods consider a sequence from different perspectives, and the fluctuations that they capture are different. The EN method focuses on l in $y(w, l)$, whereas Taylor's method focuses on w. Both characterize a text as having fluctuations amplified by a power law.

We can view Taylor analysis as the first step of EN analysis, computing $y(l)$ in formula (9.2). The EN method sums the variances despite the power dependency underlying events, and they accumulate to $y(l)$ for a particular l. This procedure corresponds to Taylor analysis, which produces α by analyzing the distribution for a particular l, as well. Therefore, EN analysis results are partly a consequence of Taylor analysis results.

The differences between these two methods can be summarized as follows (Tanaka-Ishii and Takahashi, 2021). First, Taylor analysis depends on l, whereas EN analysis does not. The overall qualitative understanding acquired by Taylor analysis, however, has been empirically reported not to depend too much on l, so this fact does not make the method insignificant. Moreover, Taylor analysis has the advantage of enabling control of the analysis by adjusting l. Second, the EN method can be used when the number of elements is small, such as W being a set of alphabetic characters. In contrast, Taylor analysis requires a distribution of elements.

Empirically, the EN method is statistically more stable; the exponent depends less on the data size. This quality derives from the method aggregating the power-law distributions underlying the elements. Consequently, the EN exponent can capture text categories only very roughly, although the method can distinguish a text from an i.i.d. process at any element level. In contrast, Taylor's method can distinguish text categories when it is applied to words.

Overall, although Taylor's method has some dependency on parameters, if the set of elements is large enough, it is a more direct analysis method. It also has the possibility of highlighting the qualitative differences between text categories. On the other hand, if the element set is limited to fewer than 100 elements, the EN method can reveal some properties of the system, but the exponent only roughly indicates the quality of a text.

9.4 Dimensions of Linguistic Fluctuation

The resulting exponents acquired through fluctuation analysis can be interpreted as a kind of dimension underlying the fluctuation of language. The exponents of the power laws described in Part II had similar values across all texts. In contrast, the exponents that we have seen in Part III can capture characteristics of different kinds of texts. We can thus interpret exponents of the latter type to represent a kind of dimension of self-similarity.

Especially, the EN exponent has an affinity with one kind of fractal dimension, i.e., the similarity dimension (Bunde and Havlin, 1996). The similarity dimension is defined for a geometrical object in a metric space. If expanding an object to make it a times larger requires b copies of the original object to fill the expanded object, then the similarity dimension is defined as follows:

$$D \equiv \log b / \log a. \tag{9.4}$$

For example, for a square whose edge length is m, doubling m gives a square $a = 2$ times larger by edge length, but filling this larger square requires $b = 4$ of the original squares. Therefore, the similarity dimension is $D = \log 4 / \log 2 = 2$. Similarly, the similarity dimension of a cube is $D = 3$. Another example is the Koch curve (Fig. 3.2), whose similarity dimension is $D = \log 4 / \log 3 = 1.262$. For a complex object, this dimension D is known not to be an integer.

A text is not a geometrical object, so its length cannot be considered in the same way as a geometrical object defined in a metric space. Nevertheless, it is also true that we often say that a text has a length. The Taylor and EN exponents then suggest a metaphorical interpretation with respect to the fractal dimension. The EN case shows that a text portion that is k times longer has k^ν times more fluctuation. Then, $D = \log k^\nu / \log k = \nu$. Similarly, although the Taylor exponent cannot be interpreted as the similarity dimension, it shows how words occurring k times more frequently fluctuate k^α times more.

As ν and α are exponents of power functions, they show the degree of self-similarity of texts, as explained in Sect. 3.3. The Taylor and EN analyses imply that a text cannot be treated as simply a concatenation of its portions. When portions are concatenated, they require further editing to unify the portions. In other words, a text requires linkages across portions to form an entirety. This generates holistic fluctuations, causing α and ν not to be 1. A text therefore has a property to amplify its fluctuations in a self-similar manner, as the Taylor and EN analyses both show. In other words, natural language texts create context by amplifying the fluctuations that already exist in a probabilistic process. In this sense, clustering phenomena can be viewed as an extension of the natural fluctuations that occur as a statistical consequence.

9.5   Relations Among Methods

This chapter ends by considering the relation between fluctuation analysis and the long-range correlation analysis of the previous chapter. The methods introduced in this chapter have a great advantage that the result for the i.i.d. case is theoretically clear. Therefore, they can judge whether fluctuations exist. Furthermore, they can analyze all the words in a text. In contrast, long-range correlation analysis is based on interval sequences, and to that end, only a subset of elements can be selected.

Nevertheless, the two methods highlight different aspects of a language sequence. Long-range correlation analysis considers the decay in the similarity of subsequences with respect to distance, whereas fluctuation analysis considers the variance of event occurrences.

The relation between the methods has been theoretically considered before (Voss, 1992; Trefán et al., 1994; Reif, 1965; Robinson, 1974). For some mathematically well-defined processes, the relations between the exponents have been mathematically clarified. Furthermore, some other mathematically defined sequences have positive long-range correlation but no fluctuation, or vice versa, as Chap. 16 will show. Natural language sequences are characterized by both being positive and having underlying phenomena that emerge as clustering behavior.

Overall, these first three chapters of Part III have analyzed the clustering behavior underlying texts. The fact that natural language texts follow power laws suggests that the clustering phenomena occur at different scales. This points to the existence of a universal characteristic of natural language: long memory. A word triggers occurrences of other words, and when it appears once again, it revives the memory of when it was used before. Part III has consistently shown how this memory of words is long. Chapter 7 showed that the probability of words returning after a long interval is not without memory. Chapter 8 examined how two sequences at a distance of a few thousand words are still correlated. Finally, this chapter has examined the scale-free aspect of the fluctuations underlying word and character occurrences in a language sequence. These chapters have shown that the long memory underlying the clustering behavior of natural language, especially that of a written text, is weak. The clustering is clearly larger than that of an i.i.d. sequence, but not by very much.

The properties described in these three chapters can also distinguish a shuffled text from the original natural language text. The next question, then, is to consider other, more sophisticated random sequences. Is it possible to construct a random sequence that produces the same set of statistical universals of natural language? If so, what different mathematical properties would it have in comparison with a natural language sequence? The answer to the question is not trivial, as Part V will examine. Before proceeding to consider that question, we still have to verify some other aspects of statistical universals. Hence, the next chapter considers the overall picture of complexity, while Part IV discusses how the statistical universals presented thus far relate to linguistic elements.

Chapter 10
Complexity

We now have a rough overview of the most important statistical universals underlying language. As a whole, is there any way to examine how complex language is? What is the characteristic underlying this complexity?

Such complexity results from the properties of the population and sequence of elements. The most important definition of complexity is, of course, that of Shannon (1951). Hence, this chapter presents a historical view of the quest to quantify the Shannon complexity of language and see how it constitutes a kind of statistical universal.

10.1 Complexity of Sequence

One simple way to consider the complexity of a sequence is via the number of possibilities. Consider a sequence of length n. If this is a binary sequence of ones and zeros, and if the ones and zeros appear equally in a uniformly random manner over length n, the number of possible sequences is obtained by repeated multiplication of 2 n times, i.e., $2 \times 2 \times \ldots \times 2$, giving 2^n possibilities.

English has 26 characters (disregarding capitals) plus a space. If every character and a space appeared uniformly randomly, then the number of possible sequences would be $27^n = 2^{(\log_2 27)n} = 2^{4.75n}$. English characters do not, of course, appear uniformly randomly. For example, "q" is almost always followed by "u." Such rules come from the conventions of a language. The number of possibilities should thus be much *less* than 27^n. The most basic exponential term consists of using 2 as the base, so let us denote this number of possibilities as 2^{hn}, where h is some constant characterizing language. Our interest now is to consider how to acquire this h. The number of possible sequences being much less than 27^n implies that $h \ll \log_2 27 = 4.75$. Then, how large could this value of h be?

© The Author(s) 2021
K. Tanaka-Ishii, *Statistical Universals of Language*, Mathematics in Mind,
https://doi.org/10.1007/978-3-030-59377-3_10

10.2   Entropy Rate

The value of h is called the entropy rate, and Claude Shannon first raised the question of how to obtain it. Historically, the answer to this question was first sought at the level of characters. Formally, let X again be a sequence of variables, $X = X_1, X_2, X_3, \ldots, X_i, \ldots, X_m$, with each variable $X_i = c \in C$, where C is a certain set of symbols, and $1 \leq i \leq m$. For entropy rate estimation in natural language, C is a set of characters, but the entropy rate can be generalized to other linguistic units such as words. For $i \leq j$, let $X_i^j \equiv X_i, X_{i+1}, \ldots, X_j$ of X, and let $P(X_i^j = c_i^j)$ denote a probability function of the subsequence X_i^j to be $c_i^j = c_i, c_{i+1}, \ldots, c_j$, with $c_i \in C$.

The Shannon entropy of a subsequence X_i^j is defined as follows (Shannon, 1948):

$$\mathrm{H}\left(X_i^j\right) \equiv -\sum_{c_i^j} P\left(X_i^j = c_i^j\right) \log P\left(X_i^j = c_i^j\right). \tag{10.1}$$

Note that the logarithmic base is 2 throughout this chapter. In the above formula, $-\log P(X_i^j = c_i^j)$ is defined as the *information* of a sequence c_i^j. The formula takes the mean of the information for different c_i^j, so H is the average information and represents the complexity underlying a sequence.

The entropy rate of X is defined in terms of H as follows (Cover and Thomas, 1991):

$$h \equiv \lim_{n \to \infty} \frac{\mathrm{H}\left(X_1^n\right)}{n}. \tag{10.2}$$

The entropy rate is thus the amount of information per element as the data length tends to infinity. There is no guarantee that a sequence has a value of h. Examples of mathematical sequences that do not have a value of h can easily be constructed (Cover and Thomas, 1991). Therefore, a natural language sequence might not have a value of h.

To estimate h according to this definition, we would have to know the probabilities of X_1^n. For language, however, it is impossible to observe $P(X_1^n)$ even for values of n that are not especially large. As the length of the context n becomes larger, counts become very small. In the case of a word sequence, even at a length of 3 or 4, they are almost zero. For example, consider a character sequence starting with "Captain Ahab is." Such sentences would seem rather likely to occur in *Moby Dick*. In fact, the counts are as follows:

"Captain": 221
"Captain Ahab": 61
"Captain Ahab is": 3

The next word is "the" in one instance and "all" in two instances. As this example shows, there is no chance to observe $P(X_1^n)$ for a large n in a real text. This does not mean, however, that the probability is zero. We could speculate that a sentence such as "Captain Ahab is a whale hunter" could appear, so the probability of "a" appearing after "is" is not zero, but this sentence does not actually appear in *Moby Dick*. This points to the fact that the possible combinations of words increase exponentially, whereas the sequences that actually appear are limited in number. Therefore, the phrases that actually appear are indeed rare among all possible phrases, and it is difficult to obtain $P(X_1^n)$ for a large n.

One way to tackle this problem is by breaking down X_1^n into subsequences. It has been proved that if X is stationary, then h exists and can be obtained as follows (Cover and Thomas, 1991):[1]

$$h = \lim_{n \to \infty} H\left(X_n | X_1^{n-1}\right). \tag{10.4}$$

This means that h can be determined from the complexity of X_n, given X_1^{n-1}. Recall that Chap. 8 examined the notion of *stationarity*, which guarantees that a stochastic process is non-changing across X_i, i.e., that the probability function $P(X_i)$ does not depend on i. Although Chap. 8 concluded that language is not stationary, for a stationary sequence, the equivalence of formula (10.2) and formula (10.4) is intuitively comprehensible; the complexity of X_1^n can be broken down to that of X_n with respect to X_1^{n-1}.

By assuming stationarity and using formula (10.4), Shannon calculated a value of $h = 1.3$ bits per character for English (Shannon, 1951). He did this by asking a subject what characters would succeed given $(n-1)$-length subsequences X_1^{n-1} of samples such as "there is no reverse on a motorcycle" and "friend of mine found this out." When the length n of a subsequence is short, it is difficult to guess what X_n should be, and the number of trials until the subject gets the correct answer will show some degree of complexity. From this idea that the number of trials will be larger for a more complex estimation, Shannon estimated h. According to Moradi et al. (1998), he conducted experiments with his wife as the only subject and obtained the above value of $h = 1.3$. Figure 10.1 shows the figure presented in his paper for his estimates of $H(X_n | X_1^{n-1})$ with respect to the text length n. Section 21.7 further explains how Shannon conducted his study.

This implies that an average of $2^{1.3} = 2.46$ possible characters follow one character. Given that the number of possibilities for a uniformly random binary sequence of one and zeros is 2, language is not much more than such a binary sequence. Shannon's study was followed by other research (Cover and King, 1978),

---

[1] Note that the conditional entropy is given in general as

$$H(X|Y) = -\sum_{x,y} P(X = x, Y = y) \log P(X = x | Y = y). \tag{10.3}$$

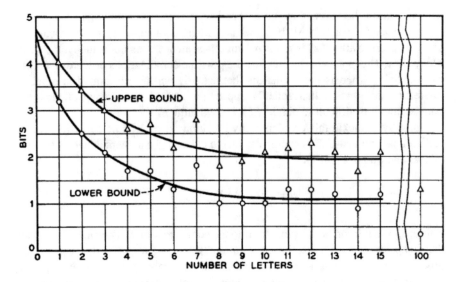

Fig. 10.1 Figure presented in Shannon (1951) to acquire h

also based on a cognitive test, that used the notion of gambling. Those authors obtained $h = 1.34$ by conducting cognitive tests with university students.

10.3   Hilberg's Ansatz

Shannon only calculated the results shown in Fig. 10.1 up to $n = 100$; his final estimate of $h = 1.3$ is thus the value at $n = 100$ (the triangle at $n = 100$). However, the figure suggests a certain decay in the h values with respect to n. If we explicitly defined a function $h(n)$, we could extrapolate $h(n)$ as $n \to \infty$ to obtain a better estimate of h than the one that Shannon derived for a limited length n. Wolfgang Hilberg followed this line of reasoning and proposed the following function $h(n)$ (Hilberg, 1990):

$$h(n) = an^{\beta-1} + h, \qquad\qquad \beta < 1, \qquad\qquad (10.5)$$

where a, β, and h are functional parameters. This first term is again a power function, and for a very large n, the first term tends to zero, i.e., $h(n) \Rightarrow h$ for $n \to \infty$.[2]

Originally, there was no term h, because Hilberg believed that $h = 0$. The notion of $h = 0$ would suggest that $2^{hn} = 2^0 = 1$, which implies that the number of

---

[2]In this book, $\Rightarrow$ indicates convergence.

possibilities would not grow with respect to n. Indeed, it might be the case that the complexity grows until a certain length n and then stops for larger values of n. For $h = 0$, however, there is a limitation with the theoretical framework. If the number of possibilities does not increase exponentially (i.e., 2^{hn}), then Shannon's information-theoretic framework gives $h = 0$. For example, if the number of possibilities grows by $2^{h'\sqrt{n}}$ for another constant h', then $h = 0$.

Contrary to Hilberg, others have presumed that $h > 0$ should hold. For example, Levy and Jaeger (2006) suggested the principle of uniform information density, meaning that h is some constant, as an aspect of rational language production. Moreover, Genzel and Charniak (2002) also considered that h should be constant within a text. Both of those works presented experimental results to suggest why h should be constant and positive. They did not disprove the possibility that $h = 0$, however, and their hypothesis of *uniform information density* is controversial. For example, Ferrer-i-Cancho et al. (2013) criticizes their claims as lacking a baseline of what is considered *uniform*. As the following chapter will show, Harris's hypothesis also provides a counterexample to information being uniform. Overall, the fundamental question of whether $h > 0$ has not yet been resolved.

There is no theoretical background for why $h(n)$ should take the form given in formula (10.5). Above all, as we do not know what kind of mathematical sequence natural language is, there is no way to induce $h(n)$ theoretically. Therefore, the function $h(n)$ is no more than an ansatz (i.e., a useful mathematical assumption). Indeed, a number of other functional forms have also been suggested (Ebeling and Nicolis, 1991; Schümann and Grassberger, 1996; Takahira et al., 2016). However, for the time being we will stick with formula (10.5), because Takahira et al. (2016) showed that it fits pretty well.

10.4 Computing Entropy Rate of Human Language

In addition to cognitive test results, we now have easy access to gigantic amounts of data. If we could compute some values of $h(n)$, then we might obtain a better estimate of h by using formula (10.5). Such an approach requires concepts developed in the field of information theory.

Two conditions are necessary to compute h for a sequence by applying an information-theoretic framework, namely, that the sequence's process is stationary and ergodic (Cover and Thomas, 1991). Section 10.2 examined the property of stationarity. Formula (10.4) also required the condition of stationarity. On the other hand, the condition of *ergodicity* guarantees that the statistical mean of data obtained by taking a large number of samples equals that obtained by taking one long sample. This ergodic characteristic is important for empirical studies in which data is finite and considered in portions, as a replacement for having $n \to \infty$. The reality, however, is that the complete mathematical premises might not be guaranteed. Nevertheless, without these two conditions, the analysis can proceed no further. Despite what we saw in the previous chapters of this part, we must provisionally accept these two assumptions for the moment, but we will return to them at the end of this chapter.

An important early work attempting to compute h was that of Brown et al. (1992). They estimated the probabilities $P(X_1^n)$ by using a trigram *language model*. A language model assumes a simple mathematical process underlying language. A shuffled text and a monkey text are the coarsest models of language. A better (but still basic) language model is the n-gram model. Section 4.4 defined an n-gram as a length of n consecutive elements in a sequence. Using this definition, the n-gram model is a kind of Markov model, as will be explained in Chap. 17.

Briefly, this model assumes that X_i is chosen only from the $n - 1$ preceding words, for a particular n $X_{i-n+1} \ldots X_{i-1}$. In other words, in a Markov model, neither elements before X_{i-n+1}, nor future elements, influence the choice of X_i. This property that an element is chosen only from the $n - 1$ preceding elements is called the *Markov property*.

Natural language exhibits the Markov property to some extent. For example, the appearance of "America" after the sequence "The United States of" is determined only by this locally preceding sequence. This is not all, however, because language has long memory, as shown in the previous chapters of Part III. Even if a language sequence could be said to have the Markov property, n would have to be *very* large for it to be apparent. Chapter 17 will describe more sophisticated language models and show that simple Markov models cannot account for the statistical universals revealed thus far.

In 1983, Brown et al. (1992) trained a trigram language model (i.e., $n = 3$) on data containing 580 million words; by analyzing test data containing 1 million words, they found that $h = 1.75$ bits per character.[3] They did not extrapolate their result, however, and 1.75 is thus merely the complexity value estimated for their data.

Subsequent studies used data compression methods to estimate h. Many compression algorithms have been devised and are commonly used in implementations such as "zip." These methods are based on the theory that h is the lower bound of any data compression applied to a sequence X (Cover and Thomas, 1991). Let $R(n)$ be a compressed, encoded bit sequence X of length n. Then, its encoding rate $r(n)$ is defined as

$$r(n) \equiv R(n)/n. \tag{10.6}$$

Certain compression algorithms are guaranteed to possess the important characteristic of being *universal*, a technical term in the field of information theory that is unrelated to statistical *universals*. Here, universality is a property of a compression method, meaning that it guarantees that $r(n) \Rightarrow h$ for $n \to \infty$, provided that the sequence X is stationary and ergodic. Many compression methods are theoretically universal, but actual implementations involve approximations to the theoretical

---

[3] See Sect. 17.2 for the concepts of language models and their training.

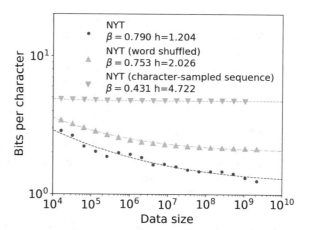

Fig. 10.2 Compression curves for the *New York Times* in the Gigaword Corpus (blue) and its shuffled (yellow middle) and monkey texts (yellow top). The horizontal axis indicates the data size n, while the vertical axis indicates $r(n)$, the encoding rate

method. For example, using an implemented compression algorithm, Bell et al. (1990) reported $h \approx 1.45$ bits per character for the collected works of Shakespeare in English.

The majority of these studies on h reported $r(n)$ for only the maximum size of the available data. In other words, they did not extrapolate $r(n)$. Our interest here, however, lies in h as $n \to \infty$. Therefore, we should calculate $r(n)$ for different values of n and extrapolate the trend by using formula (10.5).

More data points make for a better extrapolation, so even *Moby Dick* is too small for this kind of work. Thus, we will examine data from the *New York Times* part of the Gigaword Corpus. Consisting of 1.53 billion words, this corpus is one of the largest clean corpora available in English (cf. Chap. 22). The compression was conducted by a universal method called prediction by partial match, or PPM.[4] Figure 10.2 shows the results, plotted on double-logarithmic axes, for the *New York Times* and its shuffled and monkey texts.[5] The horizontal axis indicates n, while the vertical axis indicates $r(n)$ in bits. Note that this figure corresponds to Shannon's original figure shown in Fig. 10.1, but at a much larger scale.

The figure contains three plots. The blue plot indicates the result for the *New York Times* data. The two yellow plots are for the shuffled text (upward triangles, located in the middle) and the monkey text (downward triangles, located at the top). Like the *Moby Dick* monkey text, this monkey text was constructed by uniform random sampling of one character at a time from the *New York Times* for its entire length. All three plots regress pretty well to Hilberg's formula.

---

[4] The PPM code uses an n-gram-based language modeling method (Bell et al., 1990) that applies variable-length n-grams and arithmetic coding. The PPM code is guaranteed to be universal when the length of the n-gram is considered up to infinity (Ryabko, 2010). Among state-of-the-art compressors, 7-zip PPMd was used for the PPM code. PPM was used because it follows theory better than many other compression methods do, such as zip, lzh, and tar.xz (Takahira et al., 2016).

[5] I hereby thank Ryosuke Takahira and Shuntaro Takahashi for generating this figure for the purpose of this book, by reusing code used to conduct the study reported in Takahira et al. (2016).

The extrapolation of the blue plot gives $h = 1.20$, which is a little smaller than the value reported by Shannon.[6] Shannon used only 27 letters and obtained 1.3 bits per character, but the *New York Times* corpus includes both uppercase and lowercase letters and symbols, plus some common characters with diacritics. When the number of characters is larger, the value of h is empirically known to become larger, too. In the case here, however, the primary reason for the smaller value of h is the experiment's large scale, which provides better estimates of the occurrences of character subsequences. Shannon conducted his experiment with a human subject, and its size was limited. In comparison, a computer enables the use of much larger-scale data and sophisticated compression methods, and the results here were improved accordingly.

In comparison, the extrapolated entropy rates are much higher for the shuffled text ($h = 2.03$) and the monkey text ($h = 4.72$). Here, because the shuffled text includes numerous natural language words, unlike the monkey text, h is much larger for the monkey text.

On the other hand, β is the functional parameter of the first term in Hilberg's formula (10.5), in which a larger β slows the convergence to h. Therefore, the value of β indicates how hard it is for a compression algorithm to *learn to predict* a text. The results in Fig. 10.2 indicate $\beta = 0.79$ for the original text, $\beta = 0.75$ for the shuffled text, and $\beta = 0.43$ for the monkey text, so β decreases with increasing randomness. The convergence to h for natural language is indeed slow. At the same time, there is almost no difference in β between the original and shuffled texts.

The above results are similar to those obtained for large-scale corpora in other languages, especially corpora derived from collections of newspapers. In general, the Hilberg ansatz fits natural language sequences and thus may be a statistical

Fig. 10.3 β with respect to h for newspaper corpora in various languages

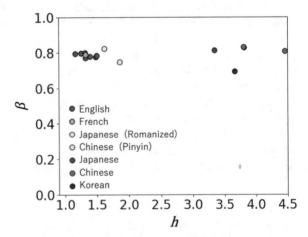

---

[6]For fitting Fig. 6.3, the least-squares method was applied (cf. Sect. 21.1). $\varepsilon = 0.0175$ for the *New York Times*, $\varepsilon = 0.00606$ for the shuffled text, and $\varepsilon = 0.00295$ for the monkey text.

universal. Figure 10.3 shows pairs of β and h for different newspaper corpora in various languages.[7] Each point represents a result for a corpus in the corresponding language. The points are in two clusters, one for alphabetic scripts (red, orange, and yellow) and the other for Japanese, Chinese, and Korean (blue, green, and purple, respectively). The values of h show large differences among the languages. The texts in alphabetic scripts show smaller h values than the texts in ideographic scripts do. This is a rather obvious result, given that the information amount per character is larger for an ideographic script than for an alphabetic script.

On the other hand, the values of β are around 0.8 for all the languages, indicating some universal aspect. We started this chapter by discussing h, but from the statistical universals perspective, the results indicate the importance of β, again highlighting the power term of formula (10.5). This finding shows that natural languages are about equally hard for the PPM compression method to learn, which implies a universal consistency of learnability across languages. A study using Lempel–Ziv compression on certain European scripts also revealed a consistent β value (Dębowski, 2015). Hilberg (1990) and a few other researchers (Crutchfield and Feldman, 2003; Ebeling and Nicolis, 1991) have hypothesized that β does not depend on a text, which would make it a language universal. In other words, this line of reasoning suggests that β is universal across languages.[8] It is especially surprising that β is consistent not only among European languages but also over totally different languages, such as Chinese and Japanese.

Whether the complexity of natural language is equal across languages has been controversial. From one perspective, the linguist Charles F. Hockett (1958) [p. 180-181] believed that the complexity is equal, and he stated this as follows:

> ... impressionistically it would seem that the total grammatical complexity of any language... is about the same as that of any other. This is not surprising, since all languages have about equally complex jobs to do ...

From the statistical universals perspective considered previously in Parts II and III, texts in any language present similar characteristics. In addition, the value β would provide one piece of evidence to support Hockett's notion.

At the same time, Fig. 10.2 shows that Hilberg's formula constitutes a good ansatz for all three plots (i.e., the plots for the natural language, shuffled, and monkey texts). The situation is similar to that shown in Part II for Zipf's law,

---

[7]This figure was adapted from Takahira et al. (2016) by applying the extrapolation function of formula (10.5) to the data listed in the first block of Table 1 in that article, which consisted only of results obtained from clean newspaper data.

[8]Whether this is true would require more fundamental research. Above all, it might not be the case that β characterizes natural language, given that a shuffled text's β was pretty close to that of natural language here. One possible path to verify the β value's universality would be to conduct a statistical test, as was done for Taylor analysis, with many sets of data as introduced in the previous section. The problem in doing so is that the texts must be very large indeed to estimate a credible β to acquire the entropy rate. At the same time, larger texts have the limitation of self-similarity as discussed in Chap. 5. Therefore, clarifying whether β is universal would require a completely different approach.

which is followed even by monkey texts. The fit to Hilberg's ansatz itself might not characterize natural language, much like Zipf's law does not. Indeed, there is also a mathematical relation between Zipf's law and the information decay presented in this chapter, and Dębowski (2020) gives an elaborate mathematical explanation of it.

10.5   Reconsidering the Question of Entropy Rate

As initiated by Shannon, the question of acquiring the true value of h has often been considered. The complexity of a sequence has been formulated via the number of possibilities for an English sequence of length n as 2^{hn}, and the value of h has been studied. This approach to understanding the complexity of language has been deemed plausible. However, as explained in this chapter, researchers have assumed various premises when they obtain h and β, yet we know from earlier chapters that those premises do not hold for language.

The previous two chapters provided some evidence that language cannot be considered stationary. Furthermore, every study reported in this chapter assumed formula (10.4), which presumes that language is a kind of Markov process. But as seen earlier in Part III, language has long memory and thus cannot be modeled only by Markov properties. In fact, Shannon himself proposed a theory to describe the distortion underlying data within the information-theoretic framework (Shannon, 1948, 1959; Berger, 1968). This distortion theory requires certain mathematical criteria to apply to sequences, but again, language sequences do not necessarily fulfill these criteria. Above all, it is probably not sufficient to consider a language sequence as being *distorted* from some simple, mathematically rigorous sequence.

Because of the limitations of working with real data and the nature of natural language, we cannot conclude with certainty whether natural language has an entropy rate at all, as mentioned at the beginning of this chapter. It then seems that estimates of h and β are no more than *some calculated values*. We must therefore question what obtained estimates of h actually represent, i.e., the significance of these values.

As long as a method is computational, h and β indicate the quality of the *language model*, as well as that of the language sequence. Indeed, the most important metric for evaluating the quality of language models in the field of computational linguistics is the *perplexity*, which is related to the value 2^h (Manning and Schütze, 1999).[9] Smaller and smaller perplexity values have been reported in the field of natural language processing, along with proposals for better language models. For example, Dai et al. (2019) reported a value corresponding to $h = 1.08$ by using state-of-the-art neural language models. Recently, a large-scale empirical report (Kaplan et al., 2020) showed how the cross entropy loss of language models

---

[9] Section 21.8 explains the perplexity in relation to the entropy rate and cross entropy.

decays by a power law as the models improve. The reasons for the decreasing perplexity lie not only in the use of larger data and better models, but also in the fact that natural language processing entirely focuses on achieving greater predictability for sequences whose mathematical behaviors are poorly understood.

Then, should we discard h from consideration as a property of language sequences? A computational method requires a model, and the resulting estimate of h indicates the quality of the model, as well as the quality of the sequence. To focus on the latter aspect, we could return to Shannon's original approach, relying not on computation but on studies with only human predictions of language. In fact, Ren et al. (2019) conducted a large-scale investigation of 700 human subjects and estimated the entropy rate of English as 1.22. However, their work was based on Shannon's original study, which again presumes stationarity. It also had some limitations due to variation among the numerous subjects.

This reality suggests that perhaps it is time to reconsider Shannon's original question. There are two different future approaches to investigate the question of h. First, we could adopt the notion of perplexity from computational linguistics, focusing on predictability. The perplexity derives from the concept of the entropy rate, but it does not presume any mathematical constraint such as the stationarity of the sequence or the (information-theoretic) universality of the method. Rather, it is a general index that represents the capability of some method of sequence prediction. Measuring this index for a very long sequence would show the prediction capabilities of a human (via a cognitive test) or a machine (via a language model), with the latter likely beating the former one day.

Second, we could adopt a more fundamental approach of studying the nature of language sequences, separately from the concept of h. To do so, we would have to investigate the true quality of a language sequence and formally describe that quality. The understanding that we gain would contribute to redefining and generalizing the concept of the entropy of language. It would also contribute to the construction of better language models, which would help achieve higher predictability. The previous chapters of Part III followed this line of inquiry by showing some possible approaches, highlighting long memory.

This chapter serves here to conclude the historical consequence of the entropy rate and show the possibility that the exponent of the power term in Hilberg's ansatz is a universal showing the predictability underlying language sequences. Furthermore, Hilberg's hypothesis plays an important role in articulating linguistic units, as we will see in the next chapter.

Part IV
Relation to Linguistic Elements and Structure

Chapter 11
Articulation of Elements

The previous two parts of this book considered statistical universals of language. Sequences were input to specific analysis methods to examine the behavior of words or characters. The resulting phenomena were studied from the two viewpoints of the poplulation and sequence. As shown by the thick rightward arrow in Fig. 1.1, Parts II and III studied language corpora to reveal the statistical universals.

From now on, the book takes the approach depicted by the leftward arrow, by considering the effects of the statistical universals on language. We will address two questions: how do statistical universals influence language? And why does language show such universal properties? Part IV especially focuses on the first question, specifically on how linguistic elements such as words and phrase structure could arise from the statistical properties of language.

To begin, this chapter shows how words, as a unit of language, can be derived via a statistical universal examined in the previous chapter. Thus far, words have been presumed to be the fundamental units of language, without considering their origin. This chapter demonstrates that words might arise from a universal decay in complexity with respect to the context length n, as explained in the previous section.

As mentioned previously, the notion of a word in this book is not the same as the one usually defined in the field of linguistics, in which a word typically is a basic unit of meaning. This chapter rather shows how a simple procedure that does not involve meaning can articulate linguistic units.

11.1 Harris's Hypothesis

In "From Phoneme to Morpheme," written in 1955 (Harris, 1955, 1968, 1988), Zellig S.Harris suggested that the linguistic boundaries of words can be detected by a count that is unrelated to meaning. Harris was one of the most important linguists of the twentieth century, and another hypothesis of his plays an important role in the next chapter, too.

© The Author(s) 2021
K. Tanaka-Ishii, *Statistical Universals of Language*, Mathematics in Mind,
https://doi.org/10.1007/978-3-030-59377-3_11

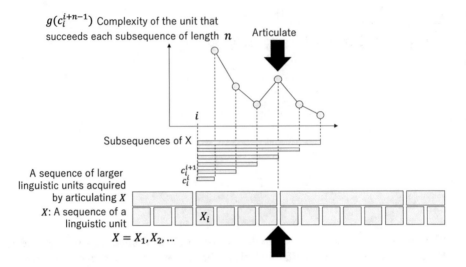

Fig. 11.1 Harris's articulation scheme

His idea about linguistic units was to observe the *successor counts* of phoneme sequences. A successor of an utterance of a given length n is the phoneme that follows the utterance. A successor count is the number of different types of successors and is obtained by going through many utterances that start with the given utterance. For example, given a short sentence "He is clever" /hiyzklevər/,[1] utterances coming after /h/ (such as "hot coffee" and "how are you") are collected, and the successor counts are measured. Then, utterances coming after /hi/ (such as "heat it" and "he is good") are collected, and the successor counts are again measured. This procedure is repeated for /hiy/, /hiyz/, /hiyzk/, and so on. Harris made the following observation (italics mine):

> When this count is made for each n of the utterance, it is found to rise and fall a number of times. If we segment the test utterance after each peak, we will find that the cuts accord *very well* with the word boundaries and *quite well* with the morpheme boundaries of that utterance.

Here, we will refer to this observation as Harris's hypothesis or scheme, as it describes a procedure to articulate words.

Figure 11.1 illustrates the idea behind Harris's hypothesis. The sequence X, at the bottom is a phoneme sequence, and this sequence forms, or articulates into, a sequence of larger elements, which are morphemes and words, as depicted by the sequence placed just above X. Harris suggested that the articulation of smaller units, i.e., phonemes, into larger ones can be done largely with statistics gathered by counting the number of different sounds that succeed a given sound, or prefix, as shown in the graph in the upper part of the figure.

---

[1]The description of pronunciation here follows that appearing in (Harris 1955).

His hypothesis filled the gap of so-called *double articulation*—first described by Martinet (1960)—which implied that language is segmented into two different kinds of elements: phonemes without meaning and morphemes with meaning. The relation between the two, however, had been unknown. Harris thus suggested that a unit of meaning can be acquired via the simple procedure of measuring the successor count.

Harris's scheme has since been studied in various contexts. Saffran (2001) argued that babies articulate words by detecting the locations of changes in the number of sound types. She conducted a cognitive experiment showing that babies indeed possess a notion similar to words. She then suggested that they detect word borders from the number of changes. This idea is essentially another version of Harris's scheme.

Moreover, Harris's scheme has been applied not only to words but also to elements of different sizes. Language has different layers of phonemes, morphemes, words, phrases, sentences, paragraphs, texts, and corpora, as shown in Fig. 1.2, and a unit in each layer consists of a sequence of elements in a lower layer. For example, a phrase consists of words.

As a concrete example, consider the compound "The United States of America." As this compound consists of a word sequence, the number of successors is counted in terms of word types rather than phonemes. The word after "The" could be almost any noun or adjective, so the successor count is very large in this case. Next, the successor count for "The United," with candidates such as "Nations" and "States," is much lower than the one for "The" but still involves multiple candidates. Finally, "The United States of" almost certainly leaves only one likely candidate.[2] This example demonstrates that the longer the prefix, the smaller the successor count. Once an atomic unit is acquired as a result of articulation, however, any phrase can follow. For the case of "The United States of America," any verb, such as "is," "wants," "negotiates," or "has," can occur, and therefore, the number of successors greatly increases. After that, the number of successors would likely start decreasing again. The number of types of successors would thus present a peak after "The United States of America." Harris's scheme articulates this compound by considering this peak as evidence of a compound border. Such articulation of compounds from a word sequence is another application of Harris's scheme, but at a different linguistic level than what Harris considered.

This explanation of how a larger unit is formed from a sequence of smaller elements shows how linguistic units are acquired recursively. The same articulation procedure can be used in different layers to acquire larger units, though the number of nested levels is limited, of course. For instance, it is unlikely that a sentence could be articulated from phrases by this scheme; moreover, as Chap. 14 will show, there is another mechanism for that purpose, namely grammar. Therefore, Harris's scheme applies only to smaller elements of language, which typically produce

---

[2]Note that there are other, far less common possibilities such as "The United States of Tara," the name of a television series, and misspelled words (see Sect. 17.3).

an inseparable, or atomic, meaning. Nonetheless, it applies recursively to a few different linguistic layers of articulated elements.

Our examination of Harris's scheme thus shows how a language sequence roughly possesses a natural capability of articulating itself. But Harris's scheme is not the only factor underlying articulation; the determination of word borders cannot be freed from various factors, including convention, chance, and social factors such as education and writing systems. Nevertheless, the effectiveness of Harris's scheme is important, and therefore, we will examine the degree to which it holds. In particular, our aim in this chapter will be to clarify what Harris meant by *very well* and *quite well* in relation to articulating words and morphemes, respectively.

11.2   Information-Theoretic Reformulation

The successor count that Harris proposed is a kind of measure of complexity.[3] Therefore, this section reformulates Harris's scheme in a modern setting. This will help clarify the scheme's connection with the universal complexity decay examined in the previous chapter.

For a sequence $X = X_1, X_2, \ldots, X_i, \ldots, X_m$, $1 \leq i \leq m$, where $X_i = c \in C$, let the probability function be $P(X_i = c)$. For Harris's scheme, C is a set of phonemes, and c is thus a phoneme. As mentioned above, however, C can also consist of elements of various layers, such as morphemes and even words. Following the notation of the previous chapter, for $i \leq j$, let X_i^j denote a sequence of variables $X_i, X_{i+1}, \ldots, X_j$, and let $c_i^j = c_i, \ldots, c_j$ be a particular sequence of X_i^j, where c_i^j is such that each element $c \in C$. Let n be the length of the sequence of previous elements used for predicting the successor element. Given the Shannon entropy introduced in formula (10.1) with a logarithmic base of 2, the complexity of the successor X_{i+n} is calculated for a particular preceding c_i^{i+n-1} by the following formula:

$$g(c_i^{i+n-1}) \equiv -\sum_{c \in C} P\left(X_{i+n} = c | X_i^{i+n-1} = c_i^{i+n-1}\right)$$

$$\times \log P\left(X_{i+n} = c | X_i^{i+n-1} = c_i^{i+n-1}\right). \tag{11.1}$$

This $g(c_i^{i+n-1})$ appears in the upper part of Fig. 11.1. The aim here is to see the fluctuation underlying $g(c_i^{i+n-1})$ for $n = 1, 2, \ldots$. In analogy to Harris's hypothesis, $g(c_i^{i+n-1})$ repeatedly falls and rises, and the locations of the peaks are the boundaries. As an example, because Harris's scheme applies to a word sequence

---

[3] Section 21.9 explains the relations of various measures of complexity through the notion of generalized entropy, including the successor count and the Shannon entropy.

like "The United States of America," $g(c_1^1)$, $g(c_1^2)$, $g(c_1^3)$, and $g(c_1^4)$ correspond to the complexities of the successors of the respective phrases "The," "The United," "The United States," and "The United States of." In that range, $g(c_i^{i+n-1})$ decreases with increasing n, but it will then rise at $g(c_1^5)$, after "The United States of America."

Harris did not provide any reason why this should be so, but the statistical universals do provide one. Chapter 10 showed that natural language has a universal decay of $H(X_n|X_1^{n-1})$ with respect to n, on average according to Hilberg's ansatz, i.e., formula (10.5). That formula described the general monotonic decay of the successor information underlying a sequence. Here, in contrast, we examine the possible successors for every actual instance,[4] and some locations show increases, which violate the general decay. Harris's scheme suggests that locations that *violate* the universal information decay are significant to humans as articulation borders. This could further imply that the statistical universal mentioned in the previous section serves as a foundation of articulation.

11.3 Accuracy of Articulation by Harris's Scheme

How accurate is Harris's scheme? Harris himself conducted a cognitive test in which he asked subjects to respond with possible successors, given a phrase. He concluded that the scheme worked *very well* for word boundaries and *quite well* for morpheme boundaries.

In the 1970s, Hafer (1974) tested how well the scheme worked on a larger scale. Their conclusion also supported Harris's hypothesis, but the scale of their experiment was still too small to evaluate the overall accuracy of Harris' hypothesis. Today, texts are processed with computers, and in the computational linguistics field, variations of ideas similar to Harris's scheme have been tested. Some of the researchers who came up with these ideas were unaware of Harris's theory, but they all proposed some variation on the "successor count." For example, instead of considering longer prefixes of length n, Kempe (1999) calculated $g(c_i^{i+n-1})$ for a fixed length of $n = 3$ across i. Kempe's model was then used by Huang and Powers (2003) for Chinese text segmentation into word sequences, though they modified it to increase the segmentation quality. Creutz and Lagus (2002) and Nobesawa et al. (1996) tested ad hoc methods for different linguistic units. Furthermore, methods similar to Harris's scheme have been used to obtain compound boundaries from a sequence of words (Frantzi and Ananiadou, 1996; Tanaka-Ishii and Ishii, 2007). The fact that so many similar proposals have been reported can be deemed evidence that articulation is a natural aspect of language sequences.

---

[4] $g(c_i^{i+n-1})$ in this chapter is designed to follow Harris's concept closely. Another way is instead to use $H(X_{i+n}|X_i^{i+n-1} = c_i^{i+n-1})$, whose relation with $H(X_n|X_1^{n-1})$ is clearer. The overall experimental result should not be different.

Surprisingly, many of the important concepts related to Harris's scheme were developed *after* his time. His scheme used the successor count, which is one way to measure the complexity, as mentioned in Sect. 11.2. Shannon presented the graph shown in Fig. 10.1 in 1965, during Harris's lifetime, whereas the Hilberg ansatz in formula (10.5) was presented in 1990. Comparing Harris's theory with the others suggests that his idea describes an aspect of the information-theoretic nature of natural language sequences.

Harris used the terms *very well* and *quite well* to indicate the degree to which his scheme worked. The rest of this chapter focuses on showing a state-of-the-art verification of his idea, following Tanaka-Ishii and Jin (2008). Instead of successor counts, we will use the Shannon complexity. Furthermore, whereas Harris considered only peak points, we will consider that all points at which $g(c_i^{i+n-1})$ violates the universal decay of $H(X_n|X_1^{n-1})$ are borders. In other words, the locations where $g(c_i^{i+n-1})$ *increases* are all regarded as word borders. This helps detect the borders of words consisting of only one element (such as "a").

To conduct a large-scale verification, we need indicators of how well a border detected by Harris's scheme corresponds with actual word and morpheme borders. A text contains actual word boundaries, while Harris's scheme produces an empirical set of word borders. The indicators should thus compare the original text and the empirical results. The standard indicators for such comparisons are the precision, recall, and f-score, defined as follows:

$$\text{precision} \equiv \frac{N_{\text{correct}}}{N_{\text{test}}}, \tag{11.2}$$

$$\text{recall} \equiv \frac{N_{\text{correct}}}{N_{\text{true}}}, \tag{11.3}$$

$$\text{f-score} \equiv \frac{2 * \text{precision} * \text{recall}}{\text{precision} + \text{recall}}, \tag{11.4}$$

where

- N_{test} is the number of empirical boundaries acquired through Harris's scheme,
- N_{correct} is the number of empirical borders that are correct, and
- N_{true} is the number of true boundaries.

Note that the precision indicates the accuracy of the method, whereas the recall indicates the coverage. For example, if "abc|def|ghi|jk" is true, with | indicating a boundary, then $N_{\text{true}} = 3$. If Harris's scheme generates empirical boundaries as "ab|cd|ef|ghi|jk," then $N_{\text{test}} = 4$. Comparing the true and empirical results shows two correct boundaries, so $N_{\text{correct}} = 2$. The precision is therefore 50% ($=\frac{2}{4}$), and the recall is 67% ($=\frac{2}{3}$). The f-score is the harmonic mean of the precision and recall, i.e., 57%.

In the study reported by Tanaka-Ishii and Jin (2008), 101 MB of data from the *Wall Street Journal* (part of the data listed in Sect. 22.2) was transformed into phoneme sequences by using the CMU Pronouncing Dictionary (The Speech Group

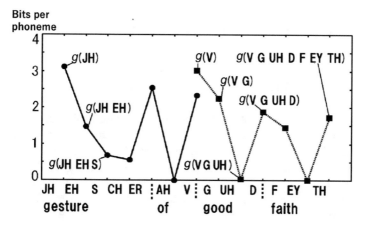

Fig. 11.2 Example of how the successor complexity serves to articulate words

at CMU, 1998). Punctuation was eliminated. After this transformation, no spaces remained to indicate word boundaries, and the data uniformly consisted of phoneme sequences. The complexity of successors was measured for 100 MB of data, which is called *training data*, and Harris's scheme was then tested with 1 MB of data (*test data*), taken from a different part of the *Wall Street Journal* source.

The articulation procedure was as follows. First, by using the 100 MB of training data, the complexities of the phonemes succeeding a subsequence of length $1 \leq n \leq 10$ were measured.[5] Then, borders were acquired for the 1 MB of test data. Let the starting point be i, where $1 \leq i < m$. For a given i, subsequences of length $n = 1, \ldots, 10$ were considered, and the question of how $g(c_i^{i+n-1})$ shifts was examined. Figure 11.1 depicts this procedure. A place where the value of $g(c_i^{i+n-1})$ increased for some n was detected as a word/morpheme border.

Figure 11.2[6] shows an actual graph of the entropy shift measured for the input phrase "JH EH S CH ER AH V G UH D F EY TH" ("gesture of good faith").[7] There are two lines for the complexity $g(c_i^{i+n-1})$, after each of the substrings starting from JH and from V. The first line (left) plots $g(\text{"JH"})$, $g(\text{"JH EH"})$ … $g(\text{"JH EH S CH ER AH V"})$. It has two increasing points, indicating that the phrase was segmented between "ER" and "AH" and after "V." The second line (right) plots

---

[5]This maximum value was chosen after testing with some larger values in the original work (Tanaka-Ishii and Jin, 2008). Language has long memory, as reported in Part III, but when limited to this specific task of articulation using Harris's scheme, $n = 10$ was deemed sufficient to obtain maximum performance.

[6]This figure appeared in Tanaka-Ishii and Jin (2008). For clarity, the figure only shows part of the experimental results. In the experiment, an entropy shift was verified starting from every phoneme for a length of 10, as mentiond in the main text.

[7]The denotation of phonemes by capital letters here follows the CMU Pronouncing Dictionary (The Speech Group at CMU, 1998).

$g(\text{"V"}) \ldots g(\text{"V G UH D F EY TH"})$. The increasing points here are between "D" and "F" and after"TH." From all the increasing points in the two lines, the following segmentation result is obtained, with | indicating a boundary:

> *gesture* | *of* | *good* | *faith*
> JH EH S CH ER | AH V | G UH D | F EY TH,

which is 100% correct word segmentation in terms of both the recall and the precision. For this example, no in-word morpheme boundaries were detected.

A similar large-scale experiment was conducted using the entire 1 MB of test data. The best f-scores were 83.6% (precision 81.0%, recall 86.3%)[8] for words, and 79.6% (precision 91.2%, recall 70.7%) for morpheme boundaries. This implies that almost half the in-word morphemes were correct ($91.2 - 81.0 = 10.2\%$, which is about half of $100 - 81 = 19.0\%$). In other words, 91% of the locations indicated as borders were correct, as either word borders or in-word morpheme borders. The precision for word borders was above 80%, meaning that articulation of four out of five borders is what Harris considered *very well*.

This large-scale verification revealed an interesting factor in this articulation: a scaling effect with respect to the size of the training corpus used to measure the complexity of the successors. Figure 11.3[9] shows the shifts in the precision and recall with respect to the corpus size. The horizontal axis indicates the corpus size on a logarithmic scale, while the vertical axis indicates the precision and recall.

We can see that morphemes have a higher precision and a lower recall compared with phonemes, which is the reverse of the case for words. Most importantly, the precision shifts almost constantly, independently of the corpus size. This result shows that Harris's hypothesis captures the basic mechanism of articulation very acutely, even with small data. For the recall, on the other hand, there are fluctuations at the beginning, but the plots both show a global, linear, increasing trend. This result indicates that an exponential amount of data is required to attain sufficient coverage.

The most frequent errors were *sectional errors*: Harris also reported these errors to be the most frequent. A common sectional error occurs when part of one morpheme is within another morpheme. For example, "all" (transliterated in phonemes as "AO L") is a morpheme, and so is "call" ("K AO L"), but because "call" includes "all," it may be segmented into "K" (thus "c") and "AO L" (thus "all"), depending on what comes before "K AO L." Another type is inter-word sectional errors. For example, "... Africa following ..." should be segmented as "AE F R IH

---

[8]The experiment involved a threshold parameter k such that candidate points at $j > i$ were regarded as borders when $g(c_i^j) - g(c_i^{j-1}) \geq k$. This k value was then varied, and f-scores were acquired for all thresholds. The value $k = 1.6$ gave the best results.

[9]This figure also appeared in Tanaka-Ishii and Jin (2008). As detailed in the original paper, the rightmost point has slightly different precision and recall values from those given in the text. The reason is that the threshold value k was varied as mentioned in the previous footnote, and this graph shows the result for another slightly different k.

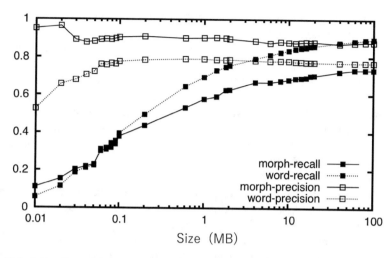

Fig. 11.3 Scale effect of corpus size: from phoneme to morpheme/word in English

K AH | F AA L OW | IH NG" but is instead segmented as "AE F R IH K | AH F | AA L OW | IH NG," because "AH F" has the same sounds as the word "of." Such sectional errors occur at both the morpheme and word levels, but more frequently at the morpheme level.

A closer analysis of Fig. 11.3 shows that all the lines are not perfectly straight but decrease slightly when the corpus size is increased: the precision lines slightly decrease linearly, while the recall lines become slightly convex. This indicates that the strength of Harris's hypothesis decreases with a larger corpus. We can speculate that proper segmentation depends on the amount of global information.

In Harris's original scheme, articulation is conducted every time anew for every instance of a phoneme sequence; therefore, the boundaries change according to the local sequence. In reality, however, we remember words, and our brains have a global memory, such as a global context or a global linguistic network of synonyms and words with similar sounds, that functions to consolidate word borders and prevent errors. By including such ideas, there are many further ways to improve the segmentation from an engineering perspective. Goldwater et al. (2009) reported one way to incorporate global optimization. Such global knowledge would suppress sectional errors, as well.

The writing system of English uses spaces as word boundaries, so English texts do not require such methods to detect word borders. The studies mentioned thus far are scientifically interesting for considering the relation between the phonetic and morphemic layers in double articulation. On the other hand, for other languages that do not clarify word boundaries by some means, methods to detect word borders have been an important engineering target. Historically, such segmentation tasks have primarily been done by using dictionaries and manually annotated segmented data, which is called a supervised method. In contrast, Harris's scheme does not rely on such additional hand-crafted data, making it a nonsupervised method. Naturally, the latter is far more difficult than the former.

Numerous verifications have been conducted on segmentation of Chinese by using an assumption similar to that of Harris's scheme, as mentioned at the beginning of this section. For comparison, the results of applying Harris's scheme to Chinese character sequences were reported in Tanaka-Ishii and Jin (2008). For the task of detecting word borders from a character sequence, the best f-score was 83.8%, with a precision of 93.9% and a recall of 75.6%. This high precision indicates that the Harris scheme works well for Chinese, too. A scale effect, similar to that shown in Fig. 11.3, was also clear. In that case, the precision was high, showing that an ideograph sequence can be segmented precisely according to Harris's hypothesis. Moreover, the recall line suggested that a further increase in the amount of data could improve the performance.

Harris's scheme therefore applies to both English and Chinese, two completely different languages. Apart from border detection of morphemes and words, as mentioned at the beginning of this subsection, there have been other attempts to articulate compounds from word sequences, as mentioned previously.

The results described here are plain results acquired with only Harris's method. Nevertheless, accuracy with values as high as 80–90% were acquired. Therefore, Harris's scheme seems to capture some nature underlying language sequences. Note again that the scheme involves no processing based on meaning, yet the resulting sets of segments largely overlap with actual words. This suggests that a language sequence is based on a structure, to articulate segments by itself. The argument of this chapter implies a possibility that the structure is partly based on a statistical universal explained in the previous chapter.

Chapter 12
Word Meaning and Value

The previous chapter suggested the possibility that linguistic units that are atomically inseparable, such as words, partly derive from statistical universals of language. Harris's hypothesis, however, only indicates how words are acquired and does not say anything about their meanings. Thus far, this book has not considered the notion of meaning at all, and the question of what is meaning is one of the most difficult in human history. It nevertheless seems that when we hear a sequence of words, we apprehend an image of its meaning. Hence, this chapter considers how the statistical universals could contribute to the meanings of words.

12.1 Meaning as Use and Distributional Semantics

As an approach to incorporating the meanings of words in our analysis, we could start by contemplating how to assign values to words. Such a valuation, in our context, must be statistically tractable. One possible approach would be to consider what a word produces through its *use*, by assuming that the value or meaning of a word lies in the way it is used. Such an approach would presume that words are identified by how they relate to other words within a sequence. In what follows we will consider the relation of statistical universals with word values acquired through use.

One way to formulate meaning as use is to mathematically describe how words distribute within a text. This book thus far has considered various distributions of words, but it has done so without any relation to meaning. The assumption of meaning as use suggests that we should study how the distribution of words could contribute to meaning. The claim that the distribution of words represents meaning is called *distributional semantics* and was in fact suggested by Harris (1954). The broadness of this claim will allow us to try different approaches to study how the distribution contributes to meaning.

© The Author(s) 2021
K. Tanaka-Ishii, *Statistical Universals of Language*, Mathematics in Mind,
https://doi.org/10.1007/978-3-030-59377-3_12

12.2 Weber–Fechner Law

One of the most basic factors in the distribution of a word is its frequency. This suggests that we consider how the frequency of a word is related to its value. One theory that could broadly relate the frequency and value is the so-called Weber–Fechner law. It is a stylized hypothesis that relates the strength of a stimulus and cognition. It is not directly about language, yet a large frequency would suggest a large stimulus.

Ernst H. Weber was a physiologist who studied the mechanisms underlying senses. The Weber–Fechner hypothesis states that the relationship between a stimulus and its perception is generally logarithmic (Fechner, 1860; Heidelberger, 1993). Mathematically, given some stimulus f and its difference Δf, their proportion should be constant:

$$\frac{\Delta f}{f} = \text{constant.} \qquad (12.1)$$

For example, a change of 10 at stimulus level 100 and a change of 100 at stimulus level 1000 should be perceived in the same way. In the case of language, suppose that a word w_a has already been heard 100 times, and that another word w_b has been heard 1000 times. Then, according to Weber, the values of hearing w_a another 10 times and w_b another 100 times should be the same. It might not be clear whether this difference forms a constant for the case of words. Nevertheless, the difference in effect between 100 and 1000 should be clear.

This law was reformulated by Gustav Fechner, a pupil of Weber's. Let g be some quantity related to the perception of a stimulus f; then,

$$g = a \log f, \qquad (12.2)$$

where a is some constant. Fechner simply integrated Weber's formulation in formula (12.1). In the case of words, f is a stimulus, the frequency of a word, while g is some perception about that word, which is proportional to the *logarithm* of the frequency f.

Previously, in Chap. 10, when the notion of information was introduced from an information-theoretic viewpoint, the logarithm of the frequency was the amount of information carried by a word (Shannon, 1951). This is a mathematical definition, and how it relates to human perception is a different question, but the Weber–Fechner viewpoint on the effect of word frequency coincides with this information-theoretic notion. Hence, we are interested in the correlation between the log frequency and the value of words.

12.3   Word Frequency and Familiarity

Consider the words "encounter" and "meeting." The word "encounter" is not as common as "meeting," although their meanings are similar. We could say that "encounter" is less familiar to us while "meeting" is more familiar. That is, we could regard the *familiarity* to be a quality of a word.

Word familiarity in this chapter means the relative ease in perception attributed to every word. In the past few decades, cognitive linguists researching language learning have attempted to measure word familiarity through experiments with human participants. In these experiments, a large number of subjects are asked to evaluate and score how familiar certain words are. The results are statistically analyzed to compute an average familiarity for every word. These studies have produced a representative familiarity list, known as the MRC database (MRC Psycholinguistic Database, 1987) in English, consisting of several thousand words with familiarity ratings. Among languages other than English, Amano's list in Japanese is a carefully worked masterpiece (Amano and Kondo, 2000) that contains about 70,000 pairs of words and corresponding ratings. In these lists, the familiarity scores range from 1.0 to 7.0, with 7.0 for the most familiar words and 1.0 for the least familiar.[1]

Such familiarity ratings have often been interpreted as a measure of the frequency of exposure to words (MRC Psycholinguistic Database, 1987), and some psychological studies have shown that a word's frequency affects its perception (Segui et al., 1982; Dupoux and Mehler, 1990; Marslen-Wilson, 1990). Another psychological study reported a relation between the word familiarity and the frequency effect in visual and auditory word recognition (Connine et al., 1990).

A basic measure would be the frequency of "exposure" to words in a text; i.e., we could relate the frequencies of words appearing in a text to familiarity ratings of those words. This would require a large corpus to measure the frequencies, because we would examine their logarithms. Furthermore, as the Amano list covers 70,000 words, the coverage of the data should be large, potentially covering the full range of daily spoken language. Even a newspaper corpus is biased, focusing on daily news content. The largest potential corpus would thus consist of web pages downloaded from the Internet. The advantage of web pages is that they contain words from every variety of source, including daily usages. Following the report in Tanaka-Ishii and

---

[1] The original ratings in the MRC list ranged from 100 to 700, so the scores were divided by 100 here for consistency with the Amano list. Furthermore, the two lists differ in that the Amano list contains only *content* words, whereas the MRC list also has some functional words. Note that functional words need not be equally familiar. For example, the familiarity of "must" is clearly higher than that of "ought." The functional words in the MRC list do not necessarily include the most frequently used functional words, such as "and" and "to."

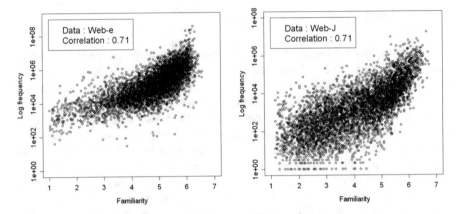

Fig. 12.1 Word familiarity rating and log frequency acquired from every word (shown by the individual points) in collections of web pages in English (left) and Japanese (right)

Terada (2011), we will examine the correlation between the frequency measured for a collection of web pages[2] downloaded from the Internet and the familiarity rating.

Figure 12.1[3] shows correlation plots of the word familiarity and the log frequency for English (left) and Japanese (right). The horizontal axes indicate the familiarity rating, whereas the vertical axes indicate the log frequency. Each point corresponds to a word that has a familiarity rating in one of the lists, while the log frequencies were computed from web page corpora.

Both plots are roughly aligned from bottom left to top right, suggesting the possibility that frequency is behind our conception of familiarity. Especially for Japanese, whose familiarity list is large and consistent, the result shows a more linear correlation between the familiarity rating and the frequency.

The legend in each graph shows the Pearson correlation coefficient, a standard measure of the correlation between two ordered lists.[4] This coefficient ranges in value from −1 (total negative correlation) to 1 (total positive correlation). For both Japanese and English, the Pearson correlation coefficient was 0.71.

---

[2]This collection was crawled from the Internet in 2006 (Tanaka-Ishii and Terada, 2011). All the markup tags were eliminated, and texts in English and Japanese were extracted. For English, 265,823,502 pages were scanned to obtain 1.9 terabytes of text data without tags. For Japanese, 12,751,271 pages were scanned to obtain 69 gigabytes of text data without tags. Chapter 22 further describes these corpora.

[3]This figure appeared in Tanaka-Ishii and Terada (2011).

[4]The figure shows the Pearson correlation coefficient because it is the most well-known such coefficient. Some readers might know that this coefficient measures the linearity of the relationship between two variables. Looking at the plots, however, we can see a nonlinearity that would be better measured by Spearman's correlation, which was 0.74 for the English data and 0.49 for the Japanese data.

The left graph of Fig. 12.1 has fewer points toward the top left but more points toward the bottom right. Such a tendency is more emphasized for results acquired from other corpora (Tanaka-Ishii and Terada, 2011). This is a natural result, because the familiarity list contains words irrespective of domains, whereas a corpus always has some bias in its content. For example, the terms "doughnut" and "coke" are familiar words in the MRC list but do not appear frequently in corpora. This suggests that frequent words tend to be familiar, whereas familiar words are not necessarily frequent for the case when the frequency count is measured from a corpus. This raises a question: what would be the conditions for a corpus to make a familiar word more frequent?

Tanaka-Ishii and Terada (2011) examined these corpus conditions to increase the correlation. The first condition lies in the size. Their work showed how the correlation coefficient increases linearly with respect to the logarithm of the corpus size. The previous chapter examined a similar dependence on the log corpus size, via Harris's hypothesis of articulation. The correlation coefficient was the highest with the largest dataset of web pages. Furthermore, when a graph like that in Fig. 12.1 is plotted for a larger corpus, the more clearly the plot falls on the region from bottom left to top right. This shows that we can alleviate the content bias, mentioned before, by increasing the corpus size.

Another condition for a corpus concerns whether it incorporates spoken words. This suggests that the domain of a corpus is probably one factor, especially because the web contains written data, whereas the spoken nature of language is also important with respect to the familiarity rating. The correlation coefficients for spoken and written data were also compared. The coefficient for the spoken data was about 0.05 better than that for the written data. This result also suggests the possibility that familiarity correlates better with the log of the frequency collected for a corpus that includes daily use of language in addition to written texts.

Overall, this analysis suggests that the logarithm of the frequency—representing the amount of information carried by a word—does correlate to some extent with the familiarity rating, when measured on a gigantic corpus of spoken, contemporary, representative language. Such evidence suggests that our conception of language could be influenced by the logarithm of the frequency, with some possible relation to the Weber–Fechner hypothesis. Because the log frequency is a key variable in Zipf's law, it might be that Zipf's law is embodied in some function of the human mind via the notion of familiarity.

12.4   Vector Representation of Words

Returning to the assumption that *the meaning of a word lies in the way it is used*, the next step is to consider the context of words. A word is used within a context, and that way of use defines the word's value. In one state-of-the-art distributional semantics, a word is represented by its neighboring words, i.e., co-occurring words.

Table 12.1 Examples of possibly co-occurring words for "coffee," "cup," and "coffee cup"

Word	Co-occurring word
Coffee	Cup, beans, Brazil, like, Blue Mountain, drink, time
Cup	Tea, whiskey, coffee, small, large, drink, china
Coffee cup	Saucer, mug, large, china, cheap, disposable, beautiful

This section and the first half of the next section follow the introductory argument in Tian (2020) in order to present an important consequence of their approach.

Table 12.1 lists some possibly co-occurring words for certain target words, following the table given in Tian (2020).[5] The target words appear in the left column, while the right column lists some possibilities for words that could appear close to the corresponding target word. For example, words that might appear around "coffee" include not only the term "cup" but also "beans," some specific origins of coffee beans, and words related to appreciating coffee. Similarly, words that might appear around "cup" include liquids for drinking and properties of cups. Furthermore, the term "coffee cup" might co-occur with terms that characterize a "coffee cup." In this way, a set of neighboring words across a corpus for a target word would represent that word.

Co-occurring words would not only possess information on the word type but also have frequency information. One way to represent the overall information is by a vector. A vector representation of a word has recently been called a distributional representation or embedding, and the acquisition of good vector representations is an important subject in computational linguistics.

Let $\vec{w}(t)$ be such a vector for a word $t \in W$. The dimension of the vector is $|W|$. All the words appearing in a corpus together with the ith element in W are represented by $\vec{w}_i(t) \equiv F(f_i)$, $i = 1, \ldots, |W|$, where f_i is the co-occurrence frequency of the ith word with t, and F is some function. In this way, the words listed in Table 12.1, for example, would be represented by $\vec{w}(\text{"coffee"})$, $\vec{w}(\text{"cup"})$, and $\vec{w}(\text{"coffee cup"})$.

Contemplating the meaning of a word by itself is difficult. Rather, the representation in vector form suggests that we can perform operations on the representation. Indeed, a vector representation is operational in that we can add and subtract vectors and calculate the similarity of two vectors.

The most important thesis of distributional semantics is the distributional hypothesis, which says that *words with a similar distribution have similar meanings* (Harris, 1954). Under the framework of vector representation, when two vectors are similar, the meanings of the words represented by those vectors are similar. Then, for the above example, it becomes an interesting question to consider whether $\vec{w}(\text{"coffee cup"})$ is close to $\vec{w}(\text{"coffee"}) \oplus \vec{w}(\text{"cup"})$, the composition of $\vec{w}(\text{"coffee"})$ and $\vec{w}(\text{"cup"})$, where the composition operator $\oplus$ denotes some compositional function of two vectors.

---

[5]The example words were changed for clarity.

12.5 Compositionality of Meaning

If we suppose that

$$\vec{w}(\text{"coffee"}) \oplus \vec{w}(\text{"cup"}) \approx \vec{w}(\text{"coffee cup"}),$$

then what does this mean?

To start with, it suggests the possibility of composition of meanings. Words are composed simply by being neighbors of each other. The term "coffee" preceding the term "cup" generates a composed image of a "coffee cup" in the mind. The notion of compositionality of meaning was coined by Frege (1892) (Beaney, 1997), but the discussion in his time remained conceptual. Today, in contrast, a vector representation is tractable and can actually be tested.

The compositional function $\oplus$ could be defined as a simple addition by devising a good function F. Then, the question would be to consider whether the following formula holds:

$$k(\vec{w}(s) + \vec{w}(t)) \approx \vec{w}(st), \tag{12.3}$$

where k is some constant. If this formula holds, then compositionality is considered to hold for the vector representation of words, which implies that meaning is compositional.

By applying this way to formulate the compositionality of meaning, Tian et al. (2017) proved an inequality that holds when the vectors meet certain conditions and the dimension $|W|$ of the vector $\vec{w}(t)$ tends to infinity, i.e., $|W| \to \infty$. Section 21.10 provides a brief summary of the proof. As a result, we have the following inequality:

$$\|\vec{w}(st) - \frac{1}{2}(\vec{w}(s) + \vec{w}(t))\|^2 \le \frac{1}{2}\left(\pi_{s\backslash t}^2 + \pi_{t\backslash s}^2 + \pi_{s\backslash t}\pi_{t\backslash s}\right). \tag{12.4}$$

Here, $s\backslash t$ denotes the state in which *word t does not appear next to word s*, and $\pi_{s\backslash t}$ then denotes the proportion of occurrences of s for which t does not appear next to s (and vice versa for $t\backslash s$ and $\pi_{t\backslash s}$). The left side of this formula is the square of the distance obtained by subtracting the left side of formula (12.3) from the right side with $k = 1/2$. From the inequality in formula (12.4), the right side gives the upper bound of compositionality. In particular, a small $\pi_{s\backslash t}$ signifies that t is likely to appear next to s whenever s appears. When both $\pi_{s\backslash t}$ and $\pi_{t\backslash s}$ are small, then the right side of formula (12.4) is small, and the meanings of s and t are better composed.

Indeed, when two words do not appear next to each other, their meanings are not composed. In our example, the terms "coffee" and "cup" are composed to form an image of a "coffee cup" because they do appear next to each other. On the other hand, a phrase such as "coffee maze" or "language cup" does not suggest any relevant image, because those words do not appear next to each other. In such cases, the right side of formula (12.4) becomes large, and the meaning is not necessarily composed.

The argument to this point has followed that of Tian (2020). This chapter must further clarify how compositionality is related to the statistical universals of language. The key to the answer lies in the conditions on (12.4). The most important condition is that the distribution of co-occurring words follows a power law.[6] When Zipf's law holds, then a power law also holds for the frequency distribution of n-grams, as Sect. 4.4 showed. Similarly, when Zipf's law holds, then a power law also holds for the frequency distribution of co-occurring words. Therefore, this condition of Tian et al. (2017) holds if Zipf's law holds. This would suggest that formula (12.4) holds for word-shuffled and monkey texts, as well.

Even if inequality (12.4) holds, however, this fact itself does not mean that the two terms are compositional. The terms are composed when the right side of formula (12.4) becomes small. This requires s and t to appear exclusively next to each other, which would mean that the terms are related tightly by co-occurrence.

In shuffled and monkey texts, s and t would not be neighbors *more* than would two words sampled from an i.i.d. sequence that follows a Zipf distribution. Under Zipf's law, two frequent words are more likely to be neighbors, because the Zipf distribution suggests a large bias among word frequencies. In natural language, however, some pairs of words appear as neighbors *even more* than they do in an i.i.d. Zipf distribution (such as a shuffled or monkey text), as we saw in Part III. In this case, the right side of formula (12.4) becomes small and compositionality holds, so that the meanings are composed.

In this formulation, the ith element of the vector $\vec{w}(t)$ is represented by a function F. Tian et al. (2017) suggested that any logarithm or square root function suffices as long as it increases slowly with respect to f_i. What this function should be mathematically for the proof to hold, however, is a different problem from how we perceive the frequency. For the latter question, the first half of this chapter considered the effect of the frequency, suggesting that F could be logarithmic, as one possibility.

Consequently, if Zipf's law holds, then inequality (12.4) holds. This is the case not only for natural language but also for any other sequence that obeys Zipf's law, including shuffled and monkey texts. We might therefore conclude that Zipf's law by itself is the basis of the compositionality described by inequality (12.4). However, the inequality only suggests the degree of compositionality existing in a text for which Zipf's law holds. For compositionality of meaning to occur between two particular words in natural language, the inequality implies that they should appear tightly next to each other among their other co-occurring words. Such phenomena would derive from the clustering behavior in a text, as seen previously in Part III. In other words, we need to study more deeply how the degree of compositionality of words is mathematically related to the clustering property of the sequence, which remains a future work.

---

[6]The proof in Tian et al. (2017) assumes multiple conditions, such as the proportionality of the co-occurrence and frequency also following a power law. Such a condition is not guaranteed to hold for natural language, in relation to the bias discussed in Chap. 5.

12.6 Statistical Universals and Meaning

Zipf himself considered that his law had something to do with meaning. He hypothesized that more frequent words tend to be more meaningful, and he thus presented the so-called *Zipf's meaning-frequency law* (Zipf, 1945). This is a power law, once again, with $f \propto k^{-a}$, where f is the frequency of a word, k is its number of meanings, and $a \approx 2$ (according to Zipf). If a rank-frequency distribution follows Zipf's law, then the number of neighboring words should follow a power function.[7] Assuming that this degree of co-occurrence has some relation with the number of meanings of a word, then it is natural for the meaning-frequency law to roughly follow a power law. In other words, Zipf's claim of power-law behavior for the meaning-frequency relation derives from Zipf's law itself.

The concrete value of a would then be determined by the assumption of a relation between the number of meanings and the degree of co-occurrence. To obtain a in this way, we would need a clear understanding of the relation between the meanings and the distribution, but it is not obvious how to gain that understanding. We could start with the number of meanings of a word appearing in the dictionary, but modern semantic modeling through the use of distributions would enable a more precise analysis. Baayen and Lieber (1996) is an example of research along this line of thought. Overall, settling the question of Zipf's meaning-frequency law will require us to study the relation between the distributions of words and their meanings, which has always been an important theme in the field of natural language processing and computational linguistics.

With this goal in mind, this chapter showed other possible influences of statistical universals on word values and meanings, which have clearer conclusions. First, the frequency of linguistic elements could contribute to human cognition of words. If the Weber–Fechner hypothesis underlies such cognition, then humans, or the human brain, might have the capability of evaluating words via the logarithm of their frequency, which suggests a possible relation to Zipf's law. Second, the compositionality of meaning could also be related to statistical universals. Under Zipf's law, the degree to which a distributional representation is compositional can be measured in terms of how words are positioned as neighbors among other co-occurring words. This fact applies to any sequence that obeys Zipf's law, including not only natural language but also a monkey text. The meanings of words are composed when they occur more exclusively next to each other within a clustering phenomenon, as described in Part III. This in turn suggests that statistical universals might contribute to the compositionality of meaning, which is a foundation of language as a system of meaning.

---

[7] As Sect. 15.2 will explain, a power distribution has the characteristic of remaining a power distribution through various kinds of transformations.

Chapter 13
Size and Frequency

Part IV thus far has examined how statistical universals might contribute to the formation of linguistic units such as words and their values. This chapter will continue to examine these units, especially in terms of the length distribution of words and compounds. Zipf analyzed this distribution and made an observation that is now known as *Zipf's law of abbreviation*, or simply *Zipf abbreviation* (Zipf, 1949) [Chapter 3]. Briefly, the law says that infrequent words are longer, and Zipf claimed that it is a universal structural property of language and is evidence of the economy underlying language.

Note, however, that with a monkey text, the greater frequency of shorter sequences is statistically trivial. Zipf abbreviation should thus be considered in relation to this fact. In comparison with a monkey text, the frequency of words in a natural language text shows many exceptions to the notion that infrequent words are long. Rather, the analysis reveals that humans use much *longer* words than are necessary, which seems to be another obvious understanding about linguistic units. The questions of what properties of linguistic units are related to their length and why they have these properties are more important, as well as indicating the fact of a correlation between the frequency and length. In particular, the answers to these questions could reconcile the holistic and constructive approaches to language.

Previously, the relation between the length of linguistic units and the frequency was considered without much comparison with a monkey text, and it was hypothesized that the relation is universal and that it is an evidence of efficiency underlying language. In comparison, this chapter provides an obvious argument from a different perspective, by showing how linguistic units are longer than those in random sequences.

© The Author(s) 2021
K. Tanaka-Ishii, *Statistical Universals of Language*, Mathematics in Mind,
https://doi.org/10.1007/978-3-030-59377-3_13

13.1 Zipf Abbreviation of Words

Zipf's law of abbreviation, or simply Zipf abbreviation, states that the longer a word is, the more infrequently it occurs. For example, in the case of the two words "meet" and "encounter," "meet" is more frequent than "encounter," and the former is shorter than the latter.

Zipf himself presented a table showing that the numbers of phonemes and syllables decrease with increasing frequency for certain words with small frequencies from 1 up to 30 occurrences in a text. Zipf thus investigated how infrequent words tend to be longer. The table, however, presented only a few examples, too few to capture the global trend.

Although a more extensive analysis is worthwhile, basing it on the number of phonemes or syllables is a limited approach, especially for English. The reason is that breaking down English words into constituent phonemes and syllables is a nontrivial problem because of the arbitrariness in the English writing system.[1] Instead, a more simplistic but computable analysis based on the word length in terms of alphabetic characters can be performed. For mathematical analysis in comparison with a monkey text, the length determines the number of occurrences, so we will examine frequency changes with respect to the length.

The left and middle graphs of Fig. 13.1 are semilog graphs of the frequency with respect to the length for *Moby Dick* and a corpus of the *New York Times*, respectively. The graphs are presented in semilog axes because of a related analytical consideration explained later in this section. The horizontal axes indicate the word length, and the vertical axes indicate the logarithm of the frequency, with the distribution displayed via box plots. The blue lines indicate the medians, the upper and lower edge of the boxes are quantiles, and the points above the boxes are outliers. In general, the word frequency decreases with the word length in both graphs.

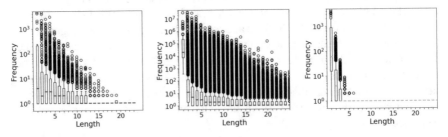

Fig. 13.1 Mean word frequency with respect to the word length for *Moby Dick* (left), the *New York Times* (middle), and a monkey text of *Moby Dick* (right)

---

[1]The procedure requires a dictionary to convert a word to a phoneme sequence. In Chap. 11, a text was transformed to phoneme sequences by using such a dictionary, but words that are not in the dictionary cannot easily be transformed into phoneme sequences.

The way the median word frequency for the *New York Times* decreases in the semilog graph suggests that it might follow a power function. Indeed, if analyzed in a double-logarithmic graph, the mean or the accumulated frequency of words of each length would present a seemingly power-like function.[2] The maximum frequencies for each length in the semilog plot, however, seem rather linear for both *Moby Dick* and the *New York Times*. On the other hand, the smallest frequency is always 1 (except for a length of 1 for the *New York Times*), and therefore, the variance of the frequencies for each length is very large, especially for shorter words.

Furthermore, it is not obvious whether Zipf abbreviation is universal across languages. In fact, Indo-European languages may roughly share a decreasing tendency similar to that of the *New York Times*. Nonetheless, in the case of *Moby Dick*, the median violates the law at a length of 2. In addition, for Chinese and Japanese the mean frequency with respect to the length would fluctuate depending on whether the length is even or odd, reflecting the grammatical characteristics underlying ideograms.

The research on Zipf abbreviation continues. Bentz and Ferrer-i-Cancho (2016) asserted that Zipf abbreviation is universal for all languages by using the Universal Declaration of Human Rights. On the other hand, by using a gigantic Google dataset, Piantadosi et al. (2011) argued that it is not the frequency that correlates well with the length but rather the information content. According to their idea, information content is the complexity of the n-grams for a term. They showed that the correlation with the length is consistently better across some Indo-European languages for the information content than for the frequency.

Zipf believed the reason for Zipf abbreviation to be some form of optimization. Bentz and Ferrer-i-Cancho (2016) also argued that this phenomenon involves optimality. However, Kanwal et al. (2017) performed a cognitive experiment showing that Zipf abbreviation only arises under certain pressures for accuracy and efficiency.

Before considering this idea of optimization, we should bear in mind that in a cruder monkey text in which all characters appear uniformly likely, Zipf abbreviation is trivial.[3] Given u characters and a word length k, with spaces to delimit words, the number of possible words is u^k. Therefore, the number of candidate words becomes large very fast with respect to k. Words must be repeated to be frequent, and this repetition is a set of probabilistic choices. On the other hand, the length of a sample text is finite. Therefore, statistically speaking, shorter words are naturally more frequent.

The right graph of Fig. 13.1 shows empirical evidence of a monkey text (sampled from *Moby Dick*). The graph indicates a rapid, consistent decrease in frequency with respect to the word length. The drop is far faster than for *Moby Dick*, and at a length

---

[2]Note that the range of lengths on the horizontal axis is too small for a logarithmic axis to reveal any useful trend, too.

[3]The corresponding graph for a shuffled text is obviously identical to that for the original natural language text.

of 4, the frequency is already almost 1. Comparison of the right graph with the left graph thus confirms that real words are much longer than for the corresponding frequency in the random sequence. Part II considered how to distinguish a monkey text from a natural language text by using the rank-frequency distribution of n-grams (Chap. 4) and the frequency density function (Chap. 6). Here, too, the difference between the first and third graphs of Fig. 13.1 shows another clear difference.

Both Miller (1957) and Mandelbrot (1953) showed that the length of a word corresponds roughly to the logarithm of its frequency rank in a crude monkey text (this idea was introduced in Sect. 4.3 and will be explained in more detail in Sects. 15.1, 21.2, and 21.11). Given Zipf's law, their results imply that the word length has a roughly linear relation with the logarithm of the frequency, which gives a linear relation between the length and the frequency on a semilog graph. This is why the graphs in Fig. 13.1 have semilog axes, with the y-axis being logarithmic but the x-axis being normal. The right graph does show a linear drop (although this linearity is difficult to see with so few points). The left and middle graphs also show a linear drop at the highest points, though the decay is faster for the median points. The mechanism underlying natural language units is different from that underlying a monkey text. Overall, the graphs show that longer words are much more frequent in natural language texts.

It is therefore unclear whether the length-frequency relation represents an actual economy underlying natural language. There are many cases in which shorter words are infrequent and long words are frequent. Moreover, the statistical comparison with a monkey text suggests the contrary, that natural language is uneconomical. Some words are arbitrarily long, and this feature characterizes language.

Words become long for multiple linguistic reasons. The alignment of characters and morphemes partly reflects the ease of pronunciation within a word. Words are articulated partly according to Harris's hypothesis, and some words are further subject to compositionality of meaning. In both cases, linguistic units are composed in a nested manner. Words are therefore not arranged randomly as in a monkey text, at any level. The length of linguistic units could be a starting point to gain a better understanding of the human factors underlying language through a study on how words are *longer* than randomly generated ones at the corresponding frequency. In other words, we might be able to study why we make language inefficient to the extent it is. Such an analysis could offer a bridge between knowledge from the holistic approach and knowledge from the constructive approach with respect to the formation of linguistic units.

13.2   Compound Length and Frequency

A similar investigation of word frequency versus length can be conducted on sequences of words. Some arrangements of words produce meaning; i.e., meaning arises in a compositional manner, as discussed in the latter half of the previous chapter. But words are often more than just compositional. For example, "green thumb" does not literally mean a green-colored thumb. Words can be even more strongly coupled; in such cases they are called compounds. One way to verify the observations in the previous section for words would be to plot the frequency of compounds with respect to length. Longer compounds obviously should be less frequent.

Verifying the length-frequency relation for compounds requires a set of compounds. Acquisition of a clean set of compounds, however, is an important problem in the field of computational linguistics and is not a trivial task. Hence, inspired by Zipf's approach, this chapter uses hyphenated compounds instead. Zipf used hyphenated compounds such as "brother-in-law" and "state-of-the-art" to conduct such analysis. Therefore, following his idea, this section considers the length-frequency relation for *hyphenated compounds*.

Measuring the frequency of hyphenated compounds requires a very large corpus, as long compounds do not readily occur in a short text. For example, *Moby Dick* has only a few hyphenated compounds. Figure 13.2[4] shows results for the *New York Times*. In this graph, similar to the previous figure, the horizontal axis indicates the lengths of hyphenated compounds in terms of the number of words, while the vertical axis indicates the frequency on a logarithmic scale[5].

The blue points show the mean frequencies for different hyphenated compounds of a given length, which are longer than one word. The plot has a general decreasing tendency but it is harder to see than in Fig. 13.1. It already fluctuates at $n = 2, 3, 4, 5$, thus leading us to question the stability of Zipf abbreviation for compounds. Viewing the graph in terms of the mean frequency thus indicates fluctuations.

Fig. 13.2 Mean frequency with respect to the length for hyphenated compounds in the *New York Times* (blue points). The black line shows the average frequency for *n*-grams with respect to the length for the *New York Times*, while the yellow line shows the same from the shuffled *New York Times*

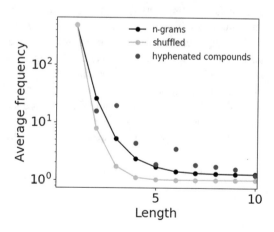

---

[4]This graph, too, is presented on semilog axes, because of Miller and Mandelbrot's theoretical analysis and the same reason mentioned for Fig. 13.1.

[5]The corpus includes some long hyphenated chunks that are sometimes doubtful to be called "compounds". Nevertheless, they are included in this analysis because they show some of the reality of hyphen usage.

The basic rationale behind this length shift of compounds is the same as the rationale given in the previous section for the shift in word length. Because we consider compounds here, an analysis based on a shuffled text would give a good comparison. We should therefore examine the length-frequency relations for hyphenated compounds in comparison with n-grams in the *New York Times* and in a shuffled text.

The black points and line in Fig. 13.2 show the mean frequency for *all n-grams* in the *New York Times*. The rank-frequency distribution of n-grams of the same length follows a power law, as shown in Fig. 4.5, whereas here the plot is of the mean frequency of each n-gram. Because the steepness of the power distribution decreases with increasing length n, the plot here has an obvious decreasing tendency with respect to n. The yellow points and line for the shuffled text also decrease rapidly as the length of the n-grams increases, thus showing that the mean frequencies are obviously smaller for longer n-grams, similarly to the previous considerations for words and compounds.

Comparing the blue and black plots, their overall behavior roughly seems pretty similar, except that many blue points appear above the black points. This result indicates that the hyphenated compounds tend to be more frequent than the ordinary n-grams. Furthermore, the yellow points are always below the black points. The implication is that the ordinary n-grams are more frequent than those of the shuffled text for each given length. This shows that in natural language, only a subset of all n-grams is more frequently used, implying that we repeat phrases.

One way to contemplate the statistical properties of a natural language text is to compare them with the properties of random sequences. While Zipf believed that Zipf abbreviation arises from economical use of language by humans, Zibf abbreviation is an obvious statistical consequence for random texts. The understanding we have gained in this chapter is rather the contrary to Zipf's reasoning and the question of Zipf abbreviation must be reformulated. The fact that humans do use longer units than those of random texts might suggest that conveying meaning requires labor. One possible reason for words being longer is that meaning is conveyed through composition, as we discussed earlier in Part IV. An investigation of the formation of linguistic units could thus provide a meeting point between the holistic understanding of language offered by the statistical universals and the understanding offered by the constructive approaches taken in linguistics.

Chapter 14
Grammatical Structure and Long Memory

The previous chapters in Part IV mainly considered the relation between the statistical universals and elements of language. This chapter instead considers the relation between long memory and the structure of sentences. Part III explained the long memory underlying language, which is characterized by clustering phenomena among linguistic events, such that words occur in a clustered manner. Such clustering phenomena have been reported in many complex systems, and their appearance in language is intuitively comprehensible as being caused by shifts in context.

Part III introduced two different but complementary ways to quantify clustering phenomena. In particular, Chap. 8 introduced long-range correlation analysis using similarity functions, which remain positive even for two subsequences separated by a long distance.

Lin and Tegmark (2017) analytically proved that a kind of grammatical sequence is associated with long-range correlation that is measurable with the mutual information function. Such grammatical structure is, of course, an important characteristic of language. Hence, this chapter considers the relation between grammar and long-range correlation. Chapter 8 confirmed that long-range correlation is barely captured with the mutual information function, implying that the long memory of natural language is *weak*. This raises the question of how to relate this weakness to grammar.

The first section introduces the simplified stochastic grammatical framework defined by Lin and Tegmark (2017). Then, the chapter discusses that framework's relation to a standard grammatical framework for natural language.

The original version of this chapter was revised. The correction to this chapter is available at
https://doi.org/10.1007/978-3-030-59377-3_23

14.1 Simple Grammatical Framework

Lin and Tegmark (2017) defined the notion of a stochastic grammar as follows. Let W be a set of elements. A sequence is generated recursively, meaning that a level-i sequence $L(i)$ expands into level $i + 1$.

Let the level-0 element be $a \in W$. An element a is generated stochastically by a probability function $P(a)$. A sequence at level $i + 1$ is generated recursively from elements at level i as follows. For every element $a \in W$ at level i, some number q of elements $b \in W$ at level $i + 1$ is generated with probability $P(b|a)$, and the resulting sequences of length q are concatenated.

For example, suppose that $q = 2$ and $W = \{1, 0\}$; Fig. 14.1 illustrates this example. A sequence is then generated stochastically via the following scheme:

Level 0: An element, either 0 or 1, is generated according to the probability functions $P(0)$ and $P(1)$, respectively, where $P(0) + P(1) = 1$. Suppose that 0 is chosen, i.e., $L(0) = [0]$ (first line in Fig. 14.1).

Level 1: A sequence is generated from the level-0 element, that is, 0. Following the probability functions $P(0|0)$ and $P(1|0)$, where $P(0|0) + P(1|0) = 1$, $q = 2$ elements are generated. For example, let these elements be $L(1) = [1, 0]$ (second line in Fig. 14.1).

Level 2: A sequence for level 2 is generated from the level-1 sequence, that is, $L(1) = [1, 0]$. For the first element, 1, this expands into a sequence of $q = 2$ elements, following $P(0|1)$ and $P(1|1)$, where $P(0|1) + P(1|1) = 1$. Suppose that this sequence is $[0, 1]$. Similarly for the second element, 0, of level 1, another sequence with $q = 2$ elements is generated, following the probability functions $P(0|0)$ and $P(1|0)$. Suppose that this sequence is $[0, 0]$. Then, $L(2) = [0, 1, 0, 0]$ is the concatenation of the sequences for each element of level 1 (third line in Fig. 14.1).

Level 3: Let each element of $L(2)$ be a. A sequence with $q = 2$ elements is generated stochastically according to $P(b|a)$, where $b \in W$, and the resulting sequences are concatenated.

...

Fig. 14.1 Simple grammatical framework of Lin and Tegmark (2017)

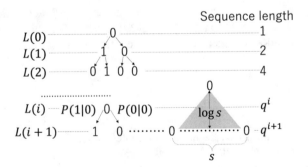

Level $i + 1$: For each element a of $L(i)$, a sequence with $q = 2$ elements is generated according to $P(b|a)$, where $b \in W$, and the resulting sequences are concatenated.

This recursive process generates a tree of depth i, where the leaf nodes form a sequence of length $|L(i)| = q^i$.

Note how the elements of lower levels influence the higher-level elements. Because the sequence develops in a tree structure, elements located next to each other at a low level influence elements far apart at a higher level. This could be the source of long memory, depending on the probability function. Suppose that the probability function to generate child nodes from a parent node is biased. If two elements b_1 and b_2, which are next to each other at a low level, happen to be different, then the sequences generated from b_1 and b_2 at a higher level become different. The sequence then becomes nonstationary. For example, suppose that $P(0|0) = 0.9$ and $P(1|1) = 0.9$, and that the elements at one level are 1 and 0. In this case, children prefer the same digit as their parents, so children below 1 are likely to prefer 1, whereas those below 0 are likely to prefer 0. Therefore, a grammar with a biased probability function can cause a nonstationary sequence.

Lin and Tegmark (2017) used the mutual information to analytically prove that the ith-level sequence should be long-range correlated. Before considering that issue in the last section, this chapter first considers the grammar introduced thus far in relation to a typical basic grammatical framework for language.

14.2 Phrase Structure Grammar

The origin of such frameworks, according to Lin and Tegmark (2017), is *Chomsky's context-free grammar* (Chomsky, 1957). As mentioned in Sect. 2.1, natural language grammar has been formulated as a set of rewriting rules, called phrase structure grammar (PSG), and context-free grammar (CFG) is one kind. One example, continuing from the previous section, is the following:

$$a \rightarrowtail b_1, b_2, \ldots, b_q, \qquad a, b_i \in W, \qquad 1 \le i \le q, \qquad (14.1)$$

where $\rightarrowtail$ indicates that the left term rewrites into the sequence on the right. In such a rewriting rule, the left term corresponds to a level-i element in the previous section, whereas the right-hand sequence corresponds to a level-$(i + 1)$ sequence.

Phrase structure grammar is a general framework using such rewriting rules. A rule may consist of multiple terms on the left and right sides. The constraints put on the elements of each side identify a hierarchy of grammar, as briefly discussed in Sect. 9.2. The above example has only one term on the left side and is called a *context-free* rewriting rule.

Note that there are several differences between the simple grammatical framework of Lin and Tegmark (2017) and CFG:

- The simple framework is probabilistic, but CFG originally was not. An updated version of CFG, however, naturally incorporates probabilities.
- CFG was originally proposed to describe sentence structure, whereas the simple framework does not have the notion of a sentence.
- The levels of recursion are not the same for all parts of the tree in CFG, whereas the simple framework has a uniform depth.
- In the simple framework, a generates constituents b_k independently across $k = 1, 2, \ldots, q$, whereas CFG describes how a expands into the subsequent level's sequence as a whole. In other words, the simple framework has no relation among the b_k, whereas the b_k can be related in CFG.
- CFG's set of words, W, has two kinds of symbols: words, and grammatical symbols such as NP (noun phrase) and VP (verb phrase).

By using CFG, we can describe the structures of real sentences. One such real example comes from a sentence in the Penn Treebank (PTB) (Marcus et al., 1994, 1993), the first large database of sentences annotated with structures (cf. Sect. 22.2):

Pierre Vinken, 61 years old, will join the board as a nonexecutive director Nov. 29.

The sentence is annotated as follows.[1]

```
( (S (NP-SBJ   (NP Pierre Vinken)
               ,
             (ADJP  (NP 61 years ) )
             old)
             , )
     (VP   will
        (VP   join
             (NP  the    board)
             (PP-CLR   as
                    (NP a nonexecutive director ) )
             (NP-TMP   Nov. 29 ) )
             .
   ) )
```

A set of parentheses indicates an expansion of one element of level i to a sequence at level $i + 1$. In the case of the previous binary example, if 1 was expanded into [1, 0], then this would be written as (1 1 0), with the first element at level i and the rest at level $i + 1$. The sequence in Fig. 14.1 up to level 2 would then be described as (0 (1 0 1) (0 0 0)).

---

[1]Symbols such as S and VP appearing in the typewriter font indicate the inner nodes of the tree and represent grammatical elements such as a *sentence* (S) and *verb phrase* (VP). The linguistic meanings of each symbol are not necessary to explain the main point of this chapter.

The real example above illustrates some rewriting rules, as follows:

```
S    ⟼    NP-SBJ VP .
VP   ⟼    will VP
VP   ⟼    join NP PP-CLR NP-TMP
```

The left side indicates an element of a lower level, which generates the sequence on the right. The three rules above are only a few of the whole set that is necessary to describe the entire structure of this particular sentence.

Note as well that the parentheses show the structure of this sentence in the form of a nested tree. The Penn Treebank has 49,208 annotated sentences of text from the *Wall Street Journal*. The annotations enable us to perform statistical analyses on those sentences. For example, the average depth is 9.05 with a standard deviation of 3.96. An important extension of this approach for large-scale, grammatically annotated data is to build a probabilistic context-free grammar, which we will examine in Chap. 17.

14.3   Long-Range Dependence in Sentences

Grammatical sequences like those above produce long-range correlation (as will be theoretically examined in the last section of this chapter), and also *long-range dependence*. Recall that both long-range correlation and long-range dependence are part of the long memory considered in this book. Long-range dependence means that elements that are separated at a long distance are related to each other. For example, the above sentence about "Pierre Vinken" indicates that the subject "Vinken" influences the verb "join," despite their separation by 5 words. This dependence can be made even longer by changing the word order:

Pierre Vinken, 61 years old, as a nonexecutive director Nov. 29, will join the board.

The distance between "Vinken" and "join" is now 11 words.

This chapter started by raising the question of how grammar relates to the long memory of natural language, which is characterized as being weak. The grammar explained thus far, however, produces all possible degrees of long memory, including both strong memory (which has deep nesting with many words showing long-range dependence across the length of an entire sentence) and weak memory (which has a limited degree of long-range dependence). If only at the level of sentences, a known linguistic universal implies how long-range dependence is suppressed, and thus provides evidence of how long memory could be weak.

That universal is one of the most well known; it is called a Greenberg universal after Joseph Greenberg, the linguist who first pointed it out (Greenberg, 1963). Greenberg identified a number of universals that reflect different properties of language; one group of universals shows potential properties in relation to word order. This book calls this theory *Greenberg's universal for word order*, and the rest of this section explains it. Since Greenberg, many researchers have elaborated his universals; representative examples are found in Dryer (1988, 1992).

This section gives a brief summary of how Greenberg's universal for word order suppresses long-range dependence. The summary requires introducing two conventional grammatical concepts used only in this section. The first is the notion of basic word order, which considers how sentences are formed from a subject (S),[2] verb (V), and object (O); any particular language has a primary order of these elements S, V, and O. In the above example, S is "Pierre Vinken," V is "join," and O is "the board." As this example indicates, the basic word order is SVO for English; it is also SVO for Chinese but SOV for Japanese.

The second concept is a dependency grammar, a model that describes a sentence structure through relations between pairs of words.[3] A dependency grammar includes the notions of *modifiers* and *heads*. When a grammatical framework generates elements b_k, $k = 1, \ldots, q$, it sets an element as a head; among the other elements at the same level, the head represents the parent. Furthermore, it functions as a *modified term*; that is, it is modified by other elements at the same level, if there are any. In our notation, the head b is one among multiple b_k, and the other b_k all modify b. We can also see this modifier-head relation in the basic word order of SVO in English: V is the head, and S and O modify the head.

Consider the case of the last VP in the example of the previous section:

```
(VP    join
       (NP   the    board)
       (PP-CLR    as
                  (NP a nonexecutive director ) )
       (NP-TMP    Nov. 29 ) )
```

The sequence ["join," NP, PP-CLR, NP-TMP] indicates *the elements at the same level under VP*; the head is "join," representing the VP (the parent of "join"), while the other elements all modify it. For example, the NP (representing "the board"), NP-TMP (representing "Nov. 29"), and PP-CLR (representing "as a nonexecutive director") all modify "join."

The question of the Greenberg universal for word order is *where* modifiers are placed with respect to the head. It indicates that the order of V and O, where O is the modifier of V, is consistent with the order of the head and the modifiers.

---

[2]The "S" of basic word order and "S" of CFG use the same symbol (with different fonts), but they are different. The former signifies the *subject*, whereas the latter signifies a *sentence*.

[3]There is a fundamental relation between a context-free grammar and a dependency grammar, via the notion of *head* explained in the main text. A sentence structure description by the former framework, however, has more information than that by the latter. A context-free grammar can almost be transformed into a dependency grammar (Buchholz, 2002), but the reverse is difficult (Kong et al., 2015; Fernández-González and Martins, 2015) because of the lack of grammatical information for the inner nodes of the tree. The latter grammar has the advantage of simplicity, and numerous treebanks have been constructed accordingly; the CoNLL database is a representative example.

For example, consider the following English expression:

the book authored by Tom.

Here, the modifiers modify the preceding heads; that is, "the book" is modified by "authored," which is modified by "by Tom." Note that this order of the head preceding its modifier is consistent with the "V then O" word order of English. On the other hand, the basic word order of Japanese is SOV. This implies that the modifier-modified order is completely reversed between English and Japanese, and this is indeed generally the case.

Consistency in word order suppresses the length of the modifier-modified dependence, making it smaller than it would be if modifiers were randomly placed before and after the head. In the above example, "the book" is modified by the term "authored" and the distance between the two terms is only one word. In contrast, suppose we rewrote the example as follows, where modifiers are placed before the head:[4]

*authored by Tom the book

Here, the head "book" and "authored" are now four words apart. The reader must decipher which words modify which other words, and it is easy to see how this longer-range dependence complicates analysis of the phrase structure.

The Greenberg universal for word order holds very roughly. In English, especially, it is easy to find exceptions such as "a charming girl" and "a blue book," where "girl" and "book" are the heads. If these phrases followed the Greenberg universal, then they would be "*a girl charming" and "*a book blue." Indeed, in other VO languages such as French, adjectives modifying nouns are usually placed after them (i.e., "une fille charmente" and "un livre bleu"). As seen here, linguistic universals have exceptions, and thus they are best called *statistical universals*, as introduced in Sect. 2.1. Linguists have thus studied reasons why certain instances do not follow universals.

Dryer (1988, 1992) suggested that such irregularities relate to the nesting characteristic of language. If a modifier is not nested, as in the case of a one-word modifier, then it often violates the modifier-modified order. One possible reason for such irregularity is again to suppress long-range dependence. In other words, a head is often modified by multiple modifiers, but reversing the order just for a one-word modifier can suppress the long-range dependence. Consider the following example:

his book authored by Tom.

If the modifier-modified order was always followed, then this would be the following:

*book his authored by Tom

---

[4]Note that * at the beginning of a linguistic example indicates that it is grammatically incorrect.

Here, "authored" is now two words apart from "book." Placing the modifier "his" before the word "book" makes the dependence ideal for all words. Therefore, not following the Greenberg universal for word order in a one-word case could avoid long-range dependence.[5]

Overall, the Greenberg universal for word order could imply that long-range dependence is *suppressed*, not enhanced. Apart from the Greenberg universal, when researchers consider why certain modifier placements are favored, they often apply the idea of shortening the range of dependence, as seen in a representative work by Hawkins (1990).

To sum up, grammar is the cause of long-range dependence, but because long memory complicates the relations among elements, there is a tendency to avoid it. The Greenberg universal for word order describes such a tendency. Hence, long memory becomes weak within a sentence, but this is only within a sentence. In Chap. 8 we saw that long memory continues to exist throughout a text despite being weak.

The means for linguistic analyses across sentences are still rather limited. A representative approach that describes global structures spanning sentences is discourse representation theory (Kamp and Reyle, 1993). Although it partly aims to describe the long-range relations existing in discourse, the main objective is to formally describe the semantics of discourse through logic. There has been no discussion of structural bias in the context of long memory. Nevertheless, natural language does possess long-range correlation throughout a text, which might lead us to wonder what kind of long memory is produced according to the theory of Lin and Tegmark's grammatical framework.

14.4 Grammatical Structure and Long-Range Correlation

Long-range dependence is, in fact, readily incorporated into a sequence generated by a simple grammatical framework. Lin and Tegmark (2017) analytically proved that a sequence generated by their simple framework has long memory captured by the mutual information function.

Their rationale can be summarized as follows. First, they showed that for a Markov process, the mutual information exponentially decays with respect to the distance s between two subsequences:

$$I(s) \propto \exp(-ks), \tag{14.2}$$

---

[5]The field of linguistics thoroughly studies such regularities and irregularities of word order, as in the database presented in Haspelmath et al. (2005), called the World Atlas of Language Structures (WALS).

for some constant k.[6] Recall that in Sect. 10.4, we defined a Markov process as one in which an element depends only on the previous n elements.

Next, Lin and Tegmark (2017) considered the distance between two elements in a sequence generated by their simple grammatical framework. Note that $P(b|a)$ in the framework itself is another relation of two words, similar to a Markov sequence, but the difference is that this relation crosses levels in the framework's tree. Let the actual distance between the two elements be s, as shown at the bottom of Fig. 14.1. Within a tree structure, two elements s apart are not generated with distance s between them. In the simple grammatical framework, elements next to each other only depend on their parents. The real distance should thus be $\log s$, the depth to a common parent. Therefore, for this simple framework,

$$I(s) \propto \exp(-k \log s) = s^{-k}, \tag{14.3}$$

indicating a power decay.

For the example condition mentioned in Sect. 14.1, with $W = \{0, 1\}$, $q = 2$, and $P(0|0) = P(1|0) = 0.9$, a million-element sequence was generated stochastically, and Fig. 14.2 shows the resulting long-range correlation. The left graph shows the result with the mutual information, similarly to Fig. 8.2. A clear long-range correlation is present, unlike the analysis of *Moby Dick* with the mutual information that was shown in Fig. 8.2. Note that this correlation would disappear, of course, with a more uniform probability function such as $P(0|0) = P(1|1) = 0.5$, whereas the more biased the probability is, such as $P(0|0) = P(1|1) = 0.9$, the stronger the correlation becomes.

Chapter 8 introduced another function for long-range correlation, the autocorrelation function. This function can also detect the long-range correlation underlying a binary sequence with biased probabilities. Despite the fact that the autocorrelation function can be used to numerically analyze a binary sequence, the original intention of Lin and Tegmark (2017) was to analyze sequences consisting of nonnumerical elements. Accordingly, the binary sequence above can be subjected to the interval analysis presented in Chap. 8, specifically by analyzing the interval sequence of 1 s. This analysis yields the graph on the right in Fig. 14.2.[7] Although the decay is far faster here than for the case of rare words in Chap. 8, a clear power tendency is present with the autocorrelation function, too.

---

[6]The original article (Lin and Tegmark, 2017) showed that $k \propto \log \phi$, where ϕ is the second largest eigenvalue of a probabilistic transition matrix M that generates the sequence, as such a matrix's largest eigenvalue is 1.

[7]$\varepsilon = 0.00565$, with a few negative values at large $s > 500$.

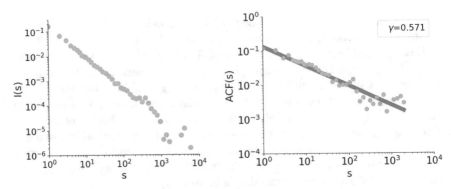

Fig. 14.2 Long-range correlation of a grammatical sequence according to the mutual information (left), and according to the autocorrelation function applied to the return interval sequence of 1 (right)

14.5   Nature of Long Memory Underlying Language

How can the results presented in this chapter characterize the long memory underlying language? To answer this question, we could start by noting what seems obvious: that grammatical structure occurring only within sentences would not produce long-range correlation. Consider an extreme text in which every sentence is completely independent, even though each one is produced by a grammatical framework. As described in Sect. 14.3, a sentence still has long-range dependence even though the Greenberg universal for word order suppresses it. However, the sentences in this extreme text are independent of each other and thus have no long-range correlation at all; we will verify this conclusion in Chap. 17.

Next, consider a less extreme text in which sentences are produced in a Markovian manner. Here, a sentence is generated only from n previous sentences, i.e., a sentence S_i is generated only from $S_{i-n}, \ldots, S_{i-1}$. Then, the long-range correlation should decay exponentially, according to the theory of Lin and Tegmark (2017), as shown by formula (14.2).

Chapter 8 showed that words are long-range correlated over at least a few thousand words. To produce such phenomena, there must be a more expansive global structure than a Markov model among sentences. Here, we could consider the grammatical construct proposed by Lin and Tegmark (2017); however, the long memory of natural language text examined in Chap. 8 was *weaker* than that of a sequence produced by their grammatical framework. The reason is that their framework guarantees long memory to appear via the mutual information function; this is not the case with natural language, as we discussed for the results shown in Fig. 8.2. At the sentence level, too, the Greenberg universal for word order shows how long-range dependence is suppressed, as discussed in the previous section. Hence, there are different pieces of evidence that consistently point to long memory being *weak*, within sentence, and across sentences, in natural language.

One way to characterize such a global structure is perhaps by the saying *more than Markov, less than a tree*. Revealing what kind of structure this is will require future work. One path lies in analyzing the behavior of rare words within the framework of Lin and Tegmark, which disregards the vocabulary distribution among elements, as presented in Part II. In other words, they considered the condition for long-range correlation to appear for a finite set. The relation between this population demography and weak long-range correlation lies as an important topic of future work to understand the nature of language.

Part V
Mathematical Models

Chapter 15
Theories Behind Zipf's Law

Part IV considered how statistical universals could inform the understanding of language that has developed in the fields of linguistics and computational linguistics. Figure 1.2 showed two different approaches to studying language, the holistic and constructive approaches. Parts II and III showed the holistic properties of natural language, while Part IV attempted to connect our understanding of those properties with the constructive understanding of language.

Here in Part V, the discussion shifts to considering why statistical universals exist. The approaches roughly consist of those from the theoretical side and those from the empirical side. The empirical approaches seek a generative process that fulfills statistical universals. The examinations with shuffled and monkey texts thus far are examples. Finding a generative process that fulfills all the statistical universals is a difficult question, as introduced in the next two chapters.

In contrast, this chapter focuses on theoretical approaches. These fall into two kinds: ideas based on optimization, and those based on statistical limit theorems. As Zipf's law is the oldest statistical universal, there have been diverse discussions on why it should hold. This chapter therefore focuses on Zipf's law and summarizes the related rationales for it. These rationales can also partly explain other statistical properties directly related to the population, as introduced in Chap. 6.

15.1 Communication Optimization

Zipf was certain that some form of economy lies behind all the language phenomena he studied. This conviction is apparent in the title of his book, *Human Behavior and the Principle of Least Effort*. Zipf compared the production of language to a daemon-bell model, as illustrated in Fig. 15.1. Here, a bell corresponds to a word.

The original version of this chapter was revised. The correction to this chapter is available at https://doi.org/10.1007/978-3-030-59377-3_23

K. Tanaka-Ishii, *Statistical Universals of Language*, Mathematics in Mind, https://doi.org/10.1007/978-3-030-59377-3_15

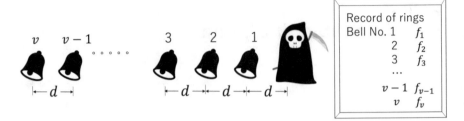

Fig. 15.1 Zipf's daemon-bell model

Every second, a daemon must ring one of v bells aligned in a row at equal distances d, until it has rung all the bells. After ringing a bell once, the daemon must return to a blackboard to record which bell it rang. For example, to ring bell No. 3, the daemon moves a distance $3d$ forward and then $3d$ backward, within a second. Zipf's model requires the daemon to ring a bell in inverse proportion to the amount of work; that is, the relation between the rank r_i of the ith frequency f_i of ringing a bell becomes $r_i \times f_i = \text{constant}$. Given the maximum frequency f_{max} for the bell that is rung the most, the total number of rings is the following:

$$f_{max} \cdot S_v = f_{max}/1 + f_{max}/2 + f_{max}/3 + \cdots + f_{max}/v, \tag{15.1}$$

where $S_v = \sum_{i=1}^{v} 1/i$. In this scheme, Zipf contemplated applying different constraints on the daemon's way of ringing the bells. For example, it might have to meet the condition that the interval lengths between respective repetitions of a bell are inversely proportional to the frequency, as in the case of the properties examined in Sect. 7.1. Zipf presumed that the daemon-bell model is a harmonic function but did not give a convincing rationale for it.

Mandelbrot (1965) gave a more thorough mathematical consideration. He showed that optimizing the average cost per unit of information implies that a scaling law underlies the rank-frequency distribution. To consider this idea, let $P(r)$ denote the probability of a word of rank r to occur,[1] and let $B(r)$ be its cost to occur. Then, the average cost is

$$\text{Cost} \equiv \sum_{r=1}^{v} P(r)B(r). \tag{15.2}$$

The information in terms of the Shannon entropy is defined by formula (10.1) as follows:

$$H \equiv -\sum_{r=1}^{v} P(r) \log_2 P(r). \tag{15.3}$$

---

[1] $P(r)$ corresponds to z introduced in Sect. 4.5, although they differ in that z represents the relative counts, whereas $P(r)$ denotes the probability under Mandelbrot's theory.

Mandelbrot then minimized the following quantity:

$$G \equiv \frac{\text{Cost}}{H},$$ (15.4)

which is the cost per unit of information. He analytically proved that minimizing G implies

$$P(r) \approx r^{-\eta}.$$ (15.5)

Section 21.11 summarizes Mandelbrot's mathematical proof to obtain formula (15.5) from formula (15.4).

Optimizing the cost per unit of information can be deemed a convincing rationale for Zipf's law. Mandelbrot's theory represents a hypothesis that a human linguistic act is optimal. The characteristic of Zipf's law is that nobody speaks or writes by aiming to produce a language sequence that obeys it. Mandelbrot's theory then suggests that humans optimize some other target function, which leads to Zipf's law. The function that Mandelbrot proposed was the cost per unit of information. Such a way to consider that *human communication is optimal*, however, is not free from controversy.

First, optimization requires some target to optimize. In the case of Mandelbrot, the optimization target was *communication*. The notion that *language is for communication* is one way to consider the objective of language, but at the same time, there is another way to consider it, namely, that language is for *thinking*, as introduced in Sect. 3.1.3. We do speak to ourselves to formulate our thoughts, and Zipf's law holds for the unfinished thoughts and non-optimized middle course of thinking. Do we optimize even when we are thinking? If so, what do we optimize? Do we not apply trial and error in looking for some representation to articulate our intention?

Second, if language is produced via optimization, it suggests that every production of a word is based on the entire discourse, even in the middle of discourse. This implies that the subsequent word must be the best choice given the whole production, including future words. Do we talk by looking ahead through the entire discourse? The act of optimization could be truly involved when the entire discourse is under human control, as in authoring a book. This presumes having the entire content in mind and the capability for multiple revisions. Yet, Zipf's law holds for a simple chat and even roughly for infant utterances. Do infants optimize as though authoring a book? Is a linguistic act not a consecutive act?

Third, another way to consider optimization is that perhaps a linguistic act is conducted according to some social convention, and this convention has become optimal through the long course of human history. It could be that people talk on such an optimal basis. With that basis, can we really say that *we optimize*? Clarifying whether we actually optimize or not in language production would require a leap in cognitive or brain studies.

Lastly, and most importantly, even if we can interpret Zipf's law in relation to mathematical optimality, whether we really do optimize, we must carefully consider the idea of optimization itself. At first glance, optimization does seem a convincing rationale that can lead to understanding.[2] Even if a phenomenon can be interpreted as optimal, however, it is not necessarily the case that the phenomenon is the outcome of optimization. Regardless of whether we optimize, it would be interesting to know that our language production can be interpreted as optimal from some mathematical perspective. Such optimality could be no more than another interpretation of phenomena. For such interpretation to be significant would require showing what it reveals further about language.

The optimization hypothesis remains compelling in some communities. Variants of Mandelbrot's scheme have been proposed that use other target optimization functions to produce either Zipf's law or some derivative properties. Ferrer-i-Cancho (2018) provided a good summary of these theories.

In Zipf's time, there seems to have been a philosophical trend to consider language from the perspective of *economy*; for instance, Martinet (1960), who was mentioned in Chap. 11, held a similar view as Zipf. That trend could explain why Zipf and Mandelbrot both applied the optimization rationale. On the other hand, Mandelbrot himself seems to have changed his mind about the economy of language, possibly along with the shifting tides of research; he suggested another mathematical way to capture what lies behind these phenomena: a limit theorem.

15.2   A Limit Theorem

Statistical properties being so universal might suggest that *any* language production inevitably shows some sort of power-law-like behavior. Statistics offers limit theorems to explain such inevitable statistical outcomes.

One such theorem is the central limit theorem. This theorem describes how the normalized error of the sum of independent draws from a distribution with a finite mean and variance tends to follow a normal distribution, as the number of events increases to infinity. Section 21.12 introduces a rough definition of the central limit theorem. It is important to note that the theorem does not depend on the kind of distribution of the information source. The occurrences of many everyday phenomena follow a normal distribution as a statistical consequence of the central limit theorem.

---

[2]One related debate involves the *variational principle* in the field of physics (Lanczos, 1949). For example, light always takes the path that reaches its destination in the shortest time, according to Fermat's principle. In relation to this principle, many diverse, complex phenomena have been interpreted under the concept of optimization.

The central limit theorem does not apply to language because the condition of finite variance (Willinger et al., 2004) is not met.[3] This means that the population distribution of words has a heavy tail. In this case, the central limit theorem indicates that the convergent distribution could be something other than a normal distribution.

One such distribution at the limit is the power distribution (Willinger et al., 2004; Stumpf and Porter, 2012) (Thurner et al., 2018, Chap. 2). Furthermore, Mandelbrot (1997) showed that power-law distributions enjoy invariant properties for various transformations. Specifically, power-law distributions are invariant under mathematical transformations such as aggregation (as in the case of the central limit theorem), mixture, maximization, and marginalization. For example, when the original events are power distributed, then their sum also follows a power distribution. The difference from the central limit theorem is that the function applied to the source is not limited only to the sum: many other functions and standard mathematical operations result in a power distribution, provided that the original information source is power distributed.

This fact implies that a power distribution has a specific characteristic of being reflexive, of returning to a power distribution for almost any transformation. Therefore, for Zipf's law and the related statistical properties to hold, there must initially be some power-law distribution. After that, any distributions that derive from it inevitably follow power laws. This implies that the utterances of a child who learns language from a power-distributed source of utterances in the environment will inevitably follow a power distribution. Stumpf and Porter (2012) characterized the abundance of power laws appearing in various phenomena as *a kind of limit theorem*.

Language already has a power distribution. Or, another view is to consider language as an aligned sequence of sounds; such a sequence already follows a power distribution, as examined in Sect. 4.3 through Miller's proof. Therefore, anyone who produces language according to the same social source should produce sequences that follow power laws. It is no wonder, then, that Zipf's law holds for *any* sample. In a sense, this Mandelbrot version of a limit theorem suggests that the universality of Zipf's law is mathematically trivial.

15.3 Significance of Statistical Universals

What, then, is important about statistical universals? Some may doubt their significance, given Mandelbrot's version of a limit theorem. This book, however, asserts a broader signification of statistical universals, indicating why they are worth

---

[3]An empirical variance can be computed, but if language is presumed to ideally follow a power law with a density function having ζ slightly below 2, as shown in Chap. 6, then the theoretical variance becomes infinite (Willinger et al., 2004).

studying. Hence, we will conclude this chapter by examining that signification from different viewpoints.

First, Mandelbrot's version of the limit theorem only explains why the rank-frequency distribution and related power laws follow power functions. As seen in Part II, real distributions deviate from power laws, and these deviations require study. As introduced in Chap. 5, there are nonnegligible biases deviating from power laws. For example, why does child-directed speech produce a convex rank-frequency distribution? Moreover, the rank-frequency distributions of characters and words show two extreme cases of exponential-like and power functions, respectively. How can we understand an intermediate rank-frequency distribution like that of Chinese characters in terms of these two distributions?

Second, the other universals of Part III are, by nature, different from the scaling law underlying the population, the only aspect to which the limit theory applies. The power law observed for long memory differs from that for population. We can understand this difference partly in terms of how a shuffled text has the scale-free property of population but does not produce any long memory. The fact of the population being scale-free therefore does not explain how elements are aligned. Long memory must have other (possibly more human) factors producing linguistic elements with certain patterns. The fluctuations of natural language, however, can be related to the basic fluctuations that occur in an i.i.d. process, as explained in Chap. 9. In this sense, clustering phenomena could be a natural consequence of another statistical phenomenon. Long memory is stable (often across a range of three decades for the values of the variables), and the exponents estimated from long-memory phenomena distinguish the kind of data. Why, then, should the Taylor exponent for a text be 0.58, for example?

Third, even if a statistical universal is mathematically trivial, its relation with our language faculty is a different question yet to be studied. After we have determined what statistical universals exist in language, the more important issue is to study how they stipulate our human language faculty and the language system in general. All the more so because they exist beyond our control, statistical universals may influence our perception of language. They could function as a precursor and *frame* our language system. Then, would not this *frame* influence what meaning is and even who we are? How do heavy tails and power laws stipulate, prescribe, or constitute the system of meaning? In an effort to answer this question, Part IV attempted to link the statistical universals with linguistic units and structure.

Fourth, language apparently has multiple statistical universals, as discussed thus far. We may reason about each scaling property by using theory, but what lies behind *all* the statistical universals seems difficult to explain. We do not know a mathematical generative mechanism that explains what language is from the statistical universal perspective. As the following two chapters will show, it is not trivial to elucidate such a mechanism.

   Stumpf and Porter (2012) questioned the validity of power laws and raised three criteria to validate analyses based on them:

1. Are the power laws statistically supported?
2. Is there a generative process that explains the power laws?
3. What genuinely new insights have been gained by having found a power law?

Here, I presume that criterion 1 has been met, given the abundant, statistically rigorous evidence reported in journal articles and conference papers.

   Regarding criterion 2, language engineering requires generative models, called language models, as a basis for computational applications that process language; the study of language models may eventually provide a possible answer to the question of this criterion. These applications include machine translation, text classification, input systems, and so forth.

   Turning to criterion 3, for language, statistical universals offer undeniable insights into the response to criterion 2. For generative models to function as language models, they should accord with the properties of language. Thus, the genuine insight drawn from the statistical universals is the very fact that they indicate the limitation of current language models. In other words, because language models are not good at reproducing all the statistical universals considered thus far (criterion 2), they have the significance (criterion 3) of motivating researchers to improve their generative models. Hence, the following two chapters will overview this state of affairs.

Chapter 16
Mathematical Generative Models

As mentioned in the previous chapter, the second criterion of Stumpf and Porter (2012) on the validity of a power law is that there should be a generative process that produces it. A rigorous mathematical methodology for our purposes is to use a generative stochastic process to produce a sequence that obeys the statistical universals. Although the generative model itself would not have to be the actual process, it should highlight the mathematical conditions necessary to produce statistical universals and suggest reasons why such statistical universals hold.

In this chapter and the next, we will examine a set of mathematical processes that have been considered to form models of language, in terms of whether the sequences they produce embody the statistical universals described in Parts II and III. Specifically, we will look at two kinds of models. The first kind, called language models, generates a sequence formed of words. Chapter 17 overviews these models. Models of the second kind, treated in this chapter, are mathematical generative models, which consist of models based on an i.i.d. process, the Simon model, and the random walk model.

16.1 Criteria for Statistical Universals

This and the following chapter will investigate whether a probabilistic generative model obeys statistical universals. A basic, rigorous way to investigate this question is to analytically examine whether a model reproduces the statistical universals. Many recent models, however, are too complex for us to use this approach. For models with this limitation, we can instead use an empirical approach. Sample sequences can be generated by a model and subjected to statistical analyses like

© The Author(s) 2021
K. Tanaka-Ishii, *Statistical Universals of Language*, Mathematics in Mind,
https://doi.org/10.1007/978-3-030-59377-3_16

those presented in Parts II and III. We must then judge whether a given sequence obeys the statistical universals, which requires a set of appropriate criteria.[1]

Part II described the statistical universals related to the population underlying a text. A sequence produced by a generative model must meet two criteria for us to conclude that it obeys the statistical universals of population:

The rank-frequency distribution   nearly follows a power law with an exponent around -1.

The vocabulary growth   follows Heaps' law with an exponent much smaller than 1.

For the first criterion, Part II showed that real texts only roughly follow this law. As for the second criterion, Chap. 6 showed that the vocabulary increases for a monkey text according to a power function with an exponent close to $\xi = 1$. This is often the case for a generative process, whereas a natural language text increases with a smaller exponent, typically around 0.7 or less. An exponent smaller than 1 is thus an important factor to determine how well a model captures the characteristics of natural language.

Part III introduced the universals related to the sequence underlying a text. In particular, Chap. 10 presented the notion of entropy rate. It functions as a general index representing both the population and the alignment of elements; furthermore, it represents the quality of language data and language models at the same time. The computed entropy rate is related to the *perplexity*,[2] an index used in the field of computational linguistics to evaluate the predictability of generative models. In contrast, our objective in this chapter is to qualitatively examine whether generative models exhibit each property of language represented by a statistical universal. To this end, the quality in question would be better grasped via more fundamental properties than the perplexity.

We will thus focus on the statistical universals presented in the other three chapters of Part III. Hence, we will say that a sequence produced by a generative model must meet the following three criteria for us to conclude that it obeys the statistical universal of long memory in a sequence:

The return interval population   of rare words deviates from an exponential distribution and instead fits a stretched exponential function.

Long-range correlation   holds for an interval sequence of rare words.

The fluctuation   is larger than the theoretical value for an i.i.d. sequence. Because the Taylor exponent is sensitive and stable, as discussed in Chap. 9, the presence of fluctuation is defined here as the Taylor exponent being larger than 0.5.

---

[1] We will mainly follow Tanaka-Ishii and Kobayashi (2018) and Takahashi and Tanaka-Ishii (2019) in qualitatively explaining the overall consequences of these approaches. The verification here was reconsidered and reproduced for the purpose of this book, making it not entirely the same as in those previous works, though it shares the same overall messages.

[2] Section 21.8 gives the precise relation.

For the first two criteria, because the models generate discrete, arbitrary elements, we will use the return analysis technique of Chaps. 7 and 8. As mentioned in Chap. 7, we will analyze the interval sequence of a fraction of the rarest words in a sequence ($1/\psi$, with $\psi = 16$). We will then examine whether the return interval population fits a stretched exponential function. As we will see, there are cases that have large deviations from the exponential distribution and do not fit the stretched exponential function. As for the second criterion, the judgment of whether there is long-range correlation is based on whether the interval sequence presents a power decay with a positive correlation. This judgment cannot be definitive, as mentioned at the end of Chap. 8, because the results often do not show a strong correlation, and the cause could be the lack of sensitivity of the autocorrelation function. Here, we will say that the criterion is not fully met unless the plotted results are narrowly distributed. Lastly, the third criterion of fluctuation can be simply judged by whether the Taylor exponent is larger than 0.5.

16.2   Independent and Identically Distributed Sequences

We have already considered the two simplest generative models introduced in Sect. 3.5: shuffled and monkey texts. Both can be regarded as independent and identically distributed (i.i.d.) sequences, in that the elements are generated one after another according to the same distribution. A monkey text is an i.i.d. sequence generated at the level of characters, produced character by character according to the character frequency distribution of the original text. On the other hand, a shuffled text is almost i.i.d. at the level of words. This approximate aspect is due to the fact that the vocabulary is identical to that of the original.

Each chapter of Parts II and III examined shuffled and monkey texts from *Moby Dick* to verify whether they followed statistical universals. Table 16.1 summarizes these and other results. The columns of the table are separated into universals related to the population and universals related to the sequence. The first row is for *Moby Dick*, which shows evidence of all the universals, as indicated by the ✓ symbol.

Table 16.1 Summary of whether mathematical models fulfill the statistical universals

	Population universals		Sequence universals		
	Rank-frequency	Vocabulary growth	Return population	Long-range correlation	Fluctuation
Moby Dick	✓	✓	✓	✓	✓
i.i.d. (shuffled)	✓	✓	×	×	×
i.i.d. (monkey)	✓	×	×	×	×
Simon model	✓	×	×	✓	×
Pitman–Yor model	✓	✓	×	×	×
Random walk on Barabasí–Albert graph	✓	✓	✓	(✓)	✓

The second row summarizes the results for the shuffled text. As the shuffled text has exactly the same population as *Moby Dick*, it obeys the two population universals. However, as indicated by the × symbols, it obeys none of the universals related to the sequence; the shuffled text thus does not meet the criteria of long memory.

The third row lists the results for the monkey text. In this case, the rank-frequency distribution almost follows Zipf's law. As explained in Chap. 4, Miller (1957) gave a mathematical proof in support of this observation. Therefore, the rank-frequency distribution cell has ✓. On the other hand, as shown in Fig. 6.3, the vocabulary grows too fast, with the exponent being almost 1. The corresponding cell thus has ×. Finally, the table shows that a monkey text does not have any of the characteristics of long memory: the mechanism to produce a monkey text cannot generate clustering phenomena.

We can probably suppose that the results for a general i.i.d. process would be similar to those of the shuffled text, at best. Hence, we are more interested in non-i.i.d. mathematical processes; the next sections examine results for non-i.i.d. mathematical sequences that present nonstationary behavior.

16.3   Simon Model and Variants

The Simon process is a mathematical generative model (Simon, 1955) that formalizes the notion of "the rich getting richer". The rank-frequency distribution of the generated sequences is analytically known to follow a power law. It has been widely used to model complex systems, including natural language. The end of this section summarizes its application in the field of computational linguistics.

The Simon model generates elements one after another over time, either by introducing a new word or by reusing a previous one (see Sect. 21.13 for a formal description of the model). The probability of selecting a new element is a, where $0 < a < 1$, while the probability of randomly selecting a previous one is $1 - a$. Suppose, for example, that the previously generated sequence is $X = [x, y, x, z, x, z]$. Then the next element will be either a new element, with probability a, or else an element sampled randomly from $X = [x, y, x, z, x, z]$. In the case of selecting a previous element, as x occurs three times and y only once, x has the probability to be chosen three times more often than y. By continuing to sample in this way, the occurrences of element x will likely increase much more than those of y will; this is why the model is analogous to "the rich getting richer."

The elements produced by this generative model are arbitrary (i.e., a random string or random number). It has been analytically proven that the rank-frequency distribution of a sequence generated with the Simon model asymptotically follows a power law (Simon, 1955; Mitzenmacher, 2003).[3] With respect to the type-token

---

[3] a must be rather small for the resulting sequence to be a model of language.

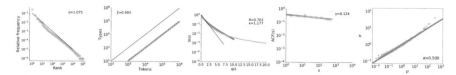

Fig. 16.1 Statistical universals for a sequence of one million elements generated by a Simon model with $a = 0.1$. From left to right, the graphs show the rank-frequency distribution (Chap. 4), vocabulary growth (Chap. 6), return interval distribution (Chap. 7), long-range correlation (Chap. 8), and Taylor analysis (Chap. 9). The yellow plots show the actual results, while the gray lines are fitted lines. The black lines in the first three graphs show lines corresponding to slopes of -1, 1, and -1, respectively

relation, the vocabulary introduction rate is constant, so it is obvious that the exponent for the vocabulary growth is 1.

The fourth row of Table 16.1 shows that the Simon model is similar to a monkey text in terms of the statistical universals for population. Moreover, the first two graphs in Fig. 16.1 show empirical results for a population sample from a Simon process.[4] The sample's length was one million, and $a = 0.1$. The exponent for the rank-frequency distribution is almost -1, although the η value is close to 1.08, because the first point at the head is too frequent, causing the value to increase. Despite this issue, the yellow plot indicates an exponent of almost -1, i.e., the slope of the black line in the graph. The vocabulary growth, in contrast, follows the theory and has an exponent of almost 1, so it is judged as *not* meeting the criterion.

As for the sequence universals, the Simon model has a biased return distribution as compared with the straight theoretical case of an exponential distribution. Whether this distribution fits a stretched exponential function, however, is questionable. The deviation seems stronger than that seen for natural language, and it does not seem to follow a stretched exponential. Thus, the Simon model does not fulfill the criterion of the return interval population.

The fourth graph shows that the sample has very strong long-range correlation. The plot barely decreases and is almost horizontal. The reason for this long-range correlation differs, however, from that for natural language. As mentioned in Sect. 4.2, any sufficiently large portions taken from different parts of a natural language text will have similar populations. On the other hand, with a Simon process, the vocabulary increases constantly. The size of the population is thus totally different between the first 10,000 words and the last 10,000 words. The uniform increase in vocabulary causes a sequence starting at the beginning and another at offset s to be strongly correlated, as both sequences include a larger number of words toward the end. Indeed, the graph shows that the correlation between two sequences at offset s barely decays at all.

---

[4]The parameters for goodness of fit were the following: rank-frequency distribution, $LL = 7.042$; vocabulary growth, $\varepsilon = 43.232$; return distribution, $\varepsilon = 0.0229$; long-range correlation, $\varepsilon = 0.0238$; Taylor analysis, $\varepsilon = 0.0121$.

The Taylor analysis in the fifth graph shows that the Simon process has an exponent of 0.5, like an i.i.d. process. This is again comprehensible, because a Simon process has no fluctuation. Some of the most frequent words present a crossover (toward the top right). This corresponds to several words located at the head of the first graph. However, we cannot conclude that the crossover occurred because of clustering phenomena.

The differences between these last two graphs and the corresponding graphs for *Moby Dick* indicate that the Simon process has a kind of long memory, but a totally different kind from that of natural language. The memory of natural language, as discussed in Part III, is based on the phenomenon of word clustering. In contrast, the Simon process has memory based on the increase in population size, and the nature of the sequence changes depending on the location. The fourth row of Table 16.1 summarizes these results. Overall, the statistical universals show that the Simon process differs from natural language.

The Simon model and its variants have been considered as models for natural language. Because the Simon model produces vocabulary growth with an exponent of 1, unlike natural language, researchers in computational linguistics turned to the Pitman–Yor process (Pitman, 2006), which generalizes the Simon process. However, although it is a generalization of the Simon process, the original mathematical definition of the Pitman–Yor process has been shown not to produce long-range memory (Tanaka-Ishii, 2018). The fifth row of Table 16.1 lists the results for this process. Because the original mathematical model behaves differently from natural language, it is doubtful that a language model made from it would obey all of the statistical universals.

Models that are based on the Pitman–Yor process by mapping elements to words have also been considered and used as language models (Goldwater et al., 2009; Teh, 2006). Indeed, Pitman–Yor language models can produce vocabulary growth similar to that of natural language. However, the resulting text behaves like the original, mathematical Pitman–Yor process, in that it lacks the sequence universals (Takahashi and Tanaka-Ishii, 2019). Moreover, there are other variants of the Simon process (Tanaka-Ishii, 2018), but as their long memory differs so greatly from that of natural language, they would have to be heavily modified for use as a language model. Accordingly, in the following chapter we will not devote any attention to language models based on the Simon or Pitman–Yor model.

16.4   Random Walk Models

None of the models considered thus far can reproduce the fluctuations underlying natural language. The previous two sections revealed the difficulty in properly modeling long memory, as compared with modeling the characteristics of population. We should thus consider how to properly model the long memory underlying language. This issue leads to the question of a fundamental model of fluctuation, and one possibility is the idea of a random walk.

The concept of a random walk is fundamental in the field of complex systems. A random walk is the theoretical basis of fluctuation analysis; moreover, the concept links the theories of long-range correlation and fluctuation analysis (Trefán et al., 1994). Note, however, that long-range correlation and fluctuation are different, as explained in Sect. 9.5, and as also seen with the Simon model, which produced a sequence that was long-range correlated but lacked fluctuation.

As treated in this book, a random walk is effectuated on a *network*, a graph structure in which nodes are connected by branches. The nodes represent elements, and the branches represent their relations. The branches can be directed and have probabilities attached to them. A "walker" starts at one arbitrary node and then walks to a neighbor node connected to the starting node. Which node the walker chooses is defined by chance, according to the probabilities attached to the nodes or branches. From the second node, the walker moves to a third node, one of the neighbors of the second, and thus the walker continues to wander the network within the connected area. A sequence is then generated from the random walk by some generation rule.[5]

Why would such a random walk produce fluctuation? The reason is that the walker visits a particular node, and then, for some time, the walker has a higher possibility to come back to that node by moving back and forth. Such revisits generate a clustering effect, thus resulting in fluctuation. Eisler et al. (2008) provided a mathematical conjecture that some random walks on a network would produce a Taylor exponent larger than 0.5. However, they did not specify the condition for a random walk to produce a sequence with a large Taylor exponent.

Inspired by Eisler *et al.*'s analysis, Tanaka-Ishii and Kobayashi (2018) investigated various representative network structures (Barabási, 2016; Thurner et al., 2018), ranging from a simple grid to graphs with very complex structures. Their goal was to determine the condition to obtain a large Taylor exponent from a random walk on a network. The study tested possible combinations of network definitions, parameters, and settings. A network sample was generated for a given combination; a walker started from a node and stochastically moved back and forth among nodes

---

[5]For example, a sequence of matching parentheses, such as "(())" or "(((())))()()()())," can be described by a random walk on a one-dimensional grid with an origin at one end and infinitely long extension toward the other end. The rule to generate the sequence is the following: "(" for a move rightward, ")" for leftward. The walker starts from the origin and must terminate the sequence by returning to the origin. Returning guarantees that the sequence has the same number of "(" and ")" characters. Note that this sequence would have long-range dependence, as the first "(" element relates to a corresponding ")" element.

The sequence does not, however, fulfill any of the statistical universals. Those for population are not applicable. The sequence has no long-range correlation as measured with either the mutual information or the autocorrelation function. The EN method (Sect. 9.1) produces an exponent of $\nu = 1$. Finally, Taylor analysis does not apply, as only two elements occur in equal number.

As seen from this example and the case of the Simon model, there are different kinds of long memory, and the characteristics of natural language represented by the term "clustering behavior" must be properly formulated.

according to the settings. The sequence of visited nodes was considered the output sequence.

The results showed that a random walk on a network can easily produce fluctuations. Even with a random walk on a two-dimensional grid (usually called a regular graph), the Taylor exponent can become far larger than 0.5. On the other hand, the population of the sequence loses its scale-free structure with such graphs, and the sequence produced no long-range correlation. This is an example of positive fluctuation with negative long-range correlation. As seen above, a Simon process generates the converse result.

The above study showed the difficulty of finding a plausible condition for a random walk on a network to obey all the statistical universals reported for natural language. Nevertheless, one network with a very limited setting could reproduce almost all the statistical universals. That limited setting was a random walk effectuated on a Barabasí–Albert (BA) graph (Barabasí and Albert, 1999). A BA graph is a scale-free graph produced by a procedure similar to the Simon process. The graph is produced by successively adding one new node at a time to the previous version of the graph. The new node is connected to k previous nodes by chance, in proportion to the number of branches that each node has. On a BA graph so constructed, the random walk is effectuated by having the walker choose neighbor nodes in proportion to their degrees (i.e., the number of branches connecting to a node).

Figure 16.2 shows the results for a sequence of one million elements produced from a BA graph with 100,000 nodes and $k = 1$.[6] The sample acquired by this random walk fulfills the criteria mentioned in Sect. 16.1 for most of the universals: the rank-frequency distribution's exponent is almost -1, the vocabulary growth has an exponent smaller than 1, the return distribution fits a stretched exponential, and the Taylor exponent is larger than 0.5. Some problems are visible for the long-range correlation, however, as the plot is rather scattered and not as stable as that for *Moby Dick*. Some values for the autocorrelation function are negative. Moreover, the decay is too fast as compared with that for *Moby Dick*. Nevertheless, a qualitative analysis

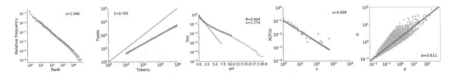

Fig. 16.2 Statistical universals for a random walk with a length of one million on a Barabasí–Albert graph with 100,000 nodes. The yellow plots represent the actual results, and the gray lines are fitted lines. The black lines in the first three graphs represent lines with slopes of -1, 1, and -1, respectively

---

[6]The parameters for goodness of fit were the following: rank-frequency distribution, $LL = 6.779$; vocabulary growth, $\varepsilon = 593.450$; return distribution, $\varepsilon = 0.0194$; long-range correlation, $\varepsilon = 0.0154$; Taylor analysis, $\varepsilon = 0.110$.

suggested partial fulfillment of this criterion, as well. The last row of Table 16.1 summarizes the results for this random walk. As long-range correlation is only reproduced in a limited manner, the ✓ symbol appears in parentheses.

This was the best result among all the many combinations of random walks tested in Tanaka-Ishii and Kobayashi (2018). Changing this setting only a little would destroy the long-range correlation. For example, in the above production scheme for the BA graph, only one branch is added for every new node. Increasing this number to two would destroy the long-range correlation and decrease the Taylor exponent, because the increased value of k would give more possibility of not returning to the same node.

This chapter has covered the representative mathematical models of language. The most important understanding here is that it is not a trivial question to find a generative scheme to reproduce all the statistical universals of natural language. Unfortunately, the following chapter, which examines the language models considered to date, reaches the same conclusion.

Chapter 17
Language Models

Because a mathematical generative model is simple, it requires only minimal thinking about the details of the underlying process, allowing us to separate the model from complex phenomena related to word production. Even if such a model obeys the statistical universals of language, however, it is not a trivial task to transform it into a generative language model that can produce a language sequence. For the purpose of language engineering, however, we need a language model.

Hence, this chapter examines representative computational language models. We will test whether they obey statistical universals, similarly to what we did in the previous chapter. In particular, the models we will look at generate a word sequence by learning from natural language data. Then, we will examine whether the sequence fulfills the criteria presented in Sect. 16.1.

The overall goal of studying such models is again to determine a generative mechanism that reproduces all of the statistical universals of natural language. However, we have seen that this is not easy to do in the case of a mathematical model; here, we will see that it is not easy to find an adequate language model either. The best candidates we have at present are neural language models, but our verification will reveal their limited effectiveness in producing statistical universals.

17.1 Language Models and Statistical Universals

Language models are studied within the field of natural language processing and computational linguistics. A language model typically outputs a probability distribution of possible words that may succeed a certain sequence of words.

On the other hand, power laws have been studied in other fields, as mentioned in Chap. 1. In particular, because a complex, large-scale system involves limit theorems, as introduced in Chap. 15, power laws are important topics in the field of statistical mechanics. Because of divisions between fields, in addition to the

© The Author(s) 2021
K. Tanaka-Ishii, *Statistical Universals of Language*, Mathematics in Mind,
https://doi.org/10.1007/978-3-030-59377-3_17

difference between holistic and constructive approaches, the statistical mechanics of language models has not been a central theme within computational linguistics. If a language model is actually to model natural language, however, it should be able to reproduce the properties of language that obey power laws.

Researchers have developed language models that learn from real data, and many of these models do approximately follow Zipf's law and have the related properties described in Part II. The Pitman–Yor model, mentioned briefly in Sect. 16.3, was the first representative model to explicitly reproduce these properties. As examined in Sect. 16.3, however, this kind of model cannot reproduce long memory.

Long memory has only recently been a consideration in the development of language models. In contrast, recent architectures in machine learning are directly related to long memory. In fact, recurrent neural network (RNN) models are known to theoretically possess memory, but they have difficulty reproducing long memory by learning from real data embodying the long-memory property. The state-of-the-art machine learning architecture, called *long short-term memory* (LSTM), was thus designed to overcome the difficulties that RNNs have in representing *long-term dependencies*, which are one kind of long memory, by elaborating the neural network architecture (Bengio et al., 1994; Pascanu et al., 2013). Today's language models, which are built from these architectures, may eventually be able to reproduce long memory. However, there has not been an established method to evaluate the quality of language models from a long-memory perspective. One possible method is to produce a long word sequence, or *pseudo-text*, with a language model and verify whether it obeys the statistical universals, as was explained in the previous chapter.

17.2   Building Language Models

Historically, roughly four kinds of models, which are qualitatively different, have been used in computational linguistics: *n*-gram models, grammatical models, Pitman–Yor models, and neural models. Section 16.3 explained the limitations of Pitman–Yor models in natural language processing; therefore, this chapter will examine the other three kinds.

Language models are constructed by learning training data consisting of a corpus. Here, *learning* means a procedure to acquire a probability distribution for outputting word sequences. Early models did so by counting the frequencies of word occurrences, whereas state-of-the-art models tune numerous underlying parameters while scanning through a text.

For training data, we will use the Penn Treebank (PTB) (Marcus et al., 1994, 1993), containing a portion of text from the *Wall Street Journal*. Chapter 14 introduced the PTB as a dataset consisting of grammatically annotated sentences. Since the PTB's construction, the *Wall Street Journal* has been used as a standard corpus for the study of computational linguistics.

The n-gram and grammatical models that we will examine in this chapter learned the entire Penn Treebank. On the other hand, the neural language models cannot handle all of the vocabulary. This is because the basic entity in a neural model is a vector, which must be finite. Therefore, neural processing of language is often conducted by collectively processing rare words as a single word. Mikolov et al. (2010) proposed to preprocess the standard Penn Treebank data to reduce the vocabulary size. This approach replaces infrequent words with <unk> and numbers with N. This is somewhat similar to processing a set of rare words as one special word, as considered in Chaps. 7 and 8. Following this convention, the neural models examined later, in Sect. 17.5, learned such preprocessed data. The problem with this cutoff appears directly in the results for the statistical universals, as discussed in that section.

All of the language models we will examine in this chapter thus learned the Penn Treebank. Each model stochastically produced a sequence of one million words. These sequences were then tested for underlying statistical universals as described in Sect. 16.1.

17.3 N-Gram Models

The n-gram model is the most fundamental language model in the field of natural language processing and computational linguistics. Such models were prevalent until the early 1990s, after which they have been used mainly as baselines for comparing the performance of more sophisticated models.

An n-gram language model is an $(n - 1)$-order Markov model. In an n-gram model, given a sequence $X_1, X_2, \ldots, X_{i-1}$, the probability of the successive element X_i is defined for a given n, $1 \leq n$, as follows:

$$P(X_i|X_1, X_2, \ldots X_{i-1}) \equiv P(X_i|X_{i-n+1}, X_{i-n+2}, \ldots, X_{i-1}). \tag{17.1}$$

This definition indicates that elements previous to X_{i-n+1} have no influence on X_i. An n-gram language model estimates the probability of X_i, given a sequence $X_{i-n+1}^{i-1} = X_{i-n+1}, \ldots, X_{i-1}$, in the form of a conditional probability. Models with $n = 2, 3$ are respectively called bigram and trigram models, as introduced in Chap. 10.

Training the model means estimating the probability distribution in formula (17.1). The simplest model obtains an estimate from empirical counts taken from a corpus. Given a set of words, W, let $\#w_i^j$ be the frequency count of $X_i = w_i, \ldots, X_j = w_j$ appearing in the corpus, where $i \leq j$ and $w_k \in W$. Then, the estimate is defined as

$$P(w_i|w_{i-n+1}^{i-1}) \equiv \frac{\#w_{i-n+1}^i}{\#w_{i-n+1}^{i-1}}. \tag{17.2}$$

However, such an estimate would cause the "zero-frequency" problem. For example, consider using 5-grams for a sequence consisting of "The United States of America," with a corpus in which that phrase appears five times and no other sequence is prefixed by "The United States of." Then,

$$P(X_i = \text{"America"} \mid X_{i-4} = \text{"The"}, X_{i-3} = \text{"United"}, X_{i-2} = \text{"States"}, X_{i-1} = \text{"of"}) = 1.$$

For all the words other than "America," the generative probability becomes 0. Such a model always produces "America" after the sequence of "The United States of," because the probability is 1.

An n-order model requires us to observe $|W|^n$ word combinations to generate the text. However, real data will have comparatively few combinations, as discussed in Sect. 10.2. The empirical counts are much fewer than the exponential number of combinations when n is large. This problem, called *sparseness*, is a major one, because probabilities cannot be acquired for unseen cases. For example, no word except "America" commonly occurs after "The United States of." Note here that the model is trained on training data, but once it is trained, it actually processes a different set of data, i.e., test data. Let us suppose that the test data includes an occurrence of "Amercia" (a misspelling) following "The United States of." Because this incorrect phrase does not appear in the training data, its probability is 0. Yet, because the phrase actually appears in the test data, its probability should not be zero. This zero-frequency problem causes numerous difficulties in processing language.

Different techniques have been proposed to tackle this problem by modifying the probabilities. Most are based on linear interpolation using counts of orders 1 to $n-1$ to acquire the n-order probability, and therefore, such techniques are often called *smoothing*, or *discounting*. Important examples of these methods include those of Stolcke (2002), Katz (1987), and Kneser and Ney (1995). We will empirically examine a sample produced by Kneser–Ney smoothing, which is known to provide one of the best n-gram models (the details are left to the original paper).

A model was thus trained with a Kneser–Ney 5-gram model on the Penn Treebank to produce a sequence of one million words. Figure 17.1 shows the results

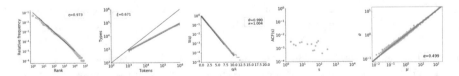

Fig. 17.1 Statistical universals for a Kneser–Ney 5-gram model trained on the Penn Treebank. The yellow plots represent the actual results, and the gray lines are fitted lines. The black lines in the first three graphs represent lines with slopes of -1, 1, and -1, respectively

Table 17.1 Summary of whether language models fulfill the statistical universals

	Population universals		Sequence universals		
	Rank-frequency	Vocabulary growth	Return population	Long-range correlation	Fluctuation
Moby Dick	✓	✓	✓	✓	✓
n-gram models	✓	✓	×	×	×
Grammatical models (applied only to sentences)	✓	✓	×	×	×
Neural models (simple RNN)	(✓)	(✓)	×	×	×
Neural models (enhanced LSTM)	(✓)	(✓)	(✓)	(✓)	(✓)

for the statistical universals.[1] The first two graphs on the left show that the model can capture the population from the corpus. It can even capture the vocabulary growth underlying language. On the other hand, the three graphs on the right show that this *n*-gram model cannot reproduce the long memory of natural language. This result is rather obvious, as an *n*-gram model only has a short memory of length *n*.

Among *n*-gram models using the simple formulation based on the Markov property definition, no variation in terms of *n* or smoothing reproduces any sign of long memory (Takahashi and Tanaka-Ishii, 2019). Moreover, an *n*-gram model is a kind of random walk; indeed, a network can be reproduced from an *n*-gram model, and a random walk on it can be regarded as a kind of probabilistic generative process. Nevertheless, a sequence produced in this way does not fulfill any of the statistical universals of long memory, either (Tanaka-Ishii and Kobayashi, 2018).

The second row of Table 17.1 summarizes the overall situation for *n*-gram models. The results show that *n*-gram models cannot reproduce long memory, which is the important characteristic of natural language.

17.4 Grammatical Models

A grammar describes the structure underlying natural language sentences. Unlike *n*-gram models, grammatical models can incorporate the long-range dependence among words. Their basis lies in the theory of phrase structure grammar, introduced in Chap. 14. The representative grammatical models used in computational linguistics are based on a probabilistic context-free grammar, and relatedly, a dependency grammar. Section 14.2 introduced a context-free grammar as a kind of phrase structure grammar. Instead of using a conditional probability function between a

---

[1]The parameters for goodness of fit were the following: rank-frequency distribution, $LL = 8.024$; vocabulary growth, $\varepsilon = 895.837$; return distribution, $\varepsilon = 0.00505$; Taylor analysis, $\varepsilon = 0.0129$. Because the sequence was not long-range correlated, the plots were not regressed to the autocorrelation function.

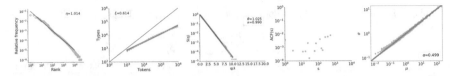

Fig. 17.2 Statistical universals for a probabilistic context-free grammar model trained on the Penn Treebank. The yellow plots represent the actual results, and the gray lines are fitted lines. The black lines in the first three graphs represent lines with slopes of -1, 1, and -1, respectively

sequence and the following element, a context-free grammar applies a probabilistic function to the inner nodes of a grammatical tree to produce the child nodes for lower levels.

Similarly to an n-gram model, it is possible to construct a probabilistic context-free grammatical language model by counting empirical occurrences in a large-scale treebank corpus. Thus, while n-gram models learn through counts of sequences, grammatical models learn through counts of subtrees. With such a trained model, a sentence can be produced probabilistically by a generative process, as introduced in Sect. 14.1. Below, we describe the analysis of a model that generated sentences totaling more than one million words.[2]

Figure 17.2 shows the results for a sample.[3] The graphs show that this model produces the statistical universals for population, but not long memory. The overall trends are the same as for the n-gram model. As mentioned in Chap. 14, Lin and Tegmark (2017) showed that a sequence produced from a simple grammatical framework (introduced in Sect. 14.1) could reproduce long memory by using the mutual information. Nevertheless, the sequence that they tested was minimal and binary, not in any form of language. Because their framework is a kind of probabilistic context-free grammar, we might see it as possible that a pseudo-text that is produced by a language model based on a probabilistic context-free grammar could also reproduce long memory. However, the last three graphs on the right in Fig. 17.2 show that this is not the case, for the reason explained at the end of Chap. 14. The samples generated by the grammatical model only at the sentence level do not produce any long memory, because no clustering occurs when sentences are produced independently.

---

[2]The sample was produced by parsing the Penn Treebank, constructing a probabilistic grammar, and generating sentences according to the grammar. No smoothing was applied in this process. The verification differed from that in a previous report (Takahashi and Tanaka-Ishii, 2019), but the overall result was the same with respect to the statistical universals of language.

[3]The sample was slightly longer than 1 million words, because the model generated units of sentences rather than words. The parameters for goodness of fit were the following: rank-frequency distribution, $LL = 7.252$; vocabulary growth, $\varepsilon = 589.667$; return distribution, $\varepsilon = 0.00635$; Taylor analysis, $\varepsilon = 0.0142$. Because the sequence was not long-range correlated, the plots were not regressed to the autocorrelation function.

Chomsky rejected Markov processes as being unable to generate long-range dependency (Chomsky, 1957), and he proposed the phrase structure grammar to introduce such dependency. If his framework is applied within a sentence, however, a text does not reproduce long memory as seen here. To reproduce long memory, a model must produce a more expansive structure that spans sentences.

The overall result for a grammatical framework based on a dependency structure, as introduced in Sect. 14.3, is the same. In so far as sentences remain independent, grammatical models cannot produce long memory. This implies that long memory is far longer than in-sentence dependencies, so the question is how to model long memory across multiple sentences.

The third row of Table 17.1 summarizes the overall situation for a probabilistic context-free grammar with respect to the statistical universals. Its failure to reproduce any sign of the long memory underlying natural language suggests the existence of a more global structure at the level of a corpus.

17.5 Neural Models

The current state-of-the-art language models are neural models. These models have been a main focus of language engineering since 2012, when they were shown to be very effective for image processing (Krizhevsky et al., 2012).

Similar to an n-gram model, a neural model predicts a subsequent element, given a context of length n. The largest difference from an n-gram model is that recent neural models are recurrent and have the capability to remember previous contexts. The influence of previous contexts lingers longer because of the recurrence mechanism. Neural models therefore seem promising with respect to long memory.

A neural architecture has layers of nodes, which are interconnected by branches. A layer represents a transformation of a vector by use of a matrix and a nonlinear function. In a recurrent network, the output of the transformation is reused as part of the subsequent input. In other words, neural processing applies multiple linear and nonlinear transformations to input vectors with recurrence to acquire an output vector. Learning of a neural architecture thus requires estimating the parameters involved in this procedure so that an input results in the required output.

To be processed by a neural network, a set of vocabulary words must be represented by a vector. We introduced an example of a vector representation in Chap. 12 by using context words. Recent neural models perform better by using sophisticated vector representations of linguistic units, which are called distributional representations or embeddings. A vector representation implies that the vocabulary must be represented by a finite set, despite a language system's characteristic of being infinite and open, as shown in Part II. A conventional way to acquire a vector representation of a vocabulary is to reduce the vocabulary by preprocessing rare words as a single special word, as mentioned in Sect. 17.2 (Mikolov et al., 2010).

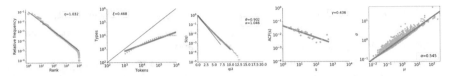

Fig. 17.3 Statistical universals for a state-of-the-art neural language model trained on the preprocessed Penn Treebank. The yellow plots represent the actual results, and the gray lines are fitted lines. The black lines in the first three graphs represent lines with slopes of -1, 1, and -1, respectively

A simple recurrent neural network (RNN) architecture was trained with the Penn Treebank after the preprocessing mentioned above. The resulting neural language model was used to stochastically generate a pseudo-text for statistical analysis. In terms of obeying the statistical universals, however, the results were worse than those for the n-gram and grammatical models. The vocabulary population roughly followed the required power laws, but with a limitation reflecting the necessity of cutting off the rarest words in the vocabulary; this limitation will be explained below. Moreover, the model produced hardly any trace of long memory: the pseudo-text's Taylor exponent was 0.50. The fourth row of Table 17.1 summarizes this situation. The results confirm the fact, mentioned in Sect. 17.1, that neural language models based on a simple RNN do not exhibit long memory.

However, such models use only the most basic neural architecture. In particular, mechanisms to enhance the performance of certain applications in machine learning, as mentioned in Sect. 17.1, have been incorporated into state-of-the-art neural language architectures. Here, Fig. 17.3 shows the results[4] for a particular state-of the art model.[5] As can be seen, this model produces the best results of the language models we have considered so far. All the graphs roughly appear to fulfill the criteria defined in Sect. 16.1. With a closer look, however, we see that the model does not completely fulfill the criterion for each statistical universal.

First, for the population, neural language models might seem to capture the scale-free tendency. A closer look reveals, however, that the tail of the rank-frequency distribution is cut off, and the vocabulary growth slows down toward the end. In fact, the simple RNN mentioned above produced a similar tendency. This is due to the nature of neural models, in that vectors are finite and rare words in the Penn Treebank are cut off, as mentioned in Sect. 17.2. The result is not sufficient because

---

[4]The parameters for goodness of fit were the following: rank-frequency distribution, $LL = 6.350$; vocabulary growth, $\varepsilon = 407.595$; return distribution, $\varepsilon = 0.00502$; long-range correlation, $\varepsilon = 0.00407$; Taylor analysis $\varepsilon = 0.0568$.

[5]This model was based on long short-term memory (LSTM) (Hochreiter and Schmidhuber, 1997) and enhanced with AWD (Merity et al., 2018), which effectively applies regularization to simple LSTM. This was combined with the techniques of continuous cache (Grave et al., 2017) and mixture of softmaxes (MoS) (Yang et al., 2018). Continuous cache is a memory augmentation architecture that computes a cache probability, which is interpolated with the AWD-LSTM probabilities. For the details, see Takahashi and Tanaka-Ishii (2019).

the statistical universals for vocabulary concern the whole population including rare words.[6] It is uncertain whether the model can capture the self-similarity underlying population.

Second, the graph of the return distribution deviates from an exponential distribution, though this deviation is not as significant as that seen for *Moby Dick* in Chap. 7. Moreover, while the fourth graph shows a power-decay tendency, there are a few negative values, and the y-axis values of the points are much lower than in the case of *Moby Dick*. There are two possible reasons for these results. The first reason is insufficient long memory, even with a state-of-the-art neural model. The second reason is the rare-word cutoff applied in the preprocessing of the Penn Treebank. To obtain the third and fourth graphs, an interval sequence consisting of only the $1/\psi$ rarest words in the whole sequence was constructed from the pseudo-text and analyzed. Because the preprocessing had already removed rare words, this interval sequence consisted of words that were *less rare* compared with the corresponding sequence of *Moby Dick*. The behavior of high-frequency words has been reported to exhibit reduced long memory and a return distribution closer to an exponential distribution (Tanaka-Ishii and Bunde, 2016). In other words, the third and fourth graphs show results for a set of more frequent words than in the case of *Moby Dick*.

The Taylor analysis graph on the far right shows an exponent of 0.55. This seems reasonable as the Penn Treebank itself has an exponent of 0.57.[7] As mentioned in Part III, Taylor analysis is advantageous because the exponent's value provides an easy way to judge whether fluctuations occur. Moreover, the Taylor exponent has been reported to be a good indicator to evaluate language models: it is correlated with the perplexity (Takahashi and Tanaka-Ishii, 2019), which is a common index for evaluating language models in general, as mentioned in Chaps. 10 and 16.

Overall, state-of-the-art neural language models can potentially reproduce the statistical universals to some extent. Our verification still required cutting off rare words, and we must interpret the results accordingly. The last row of Table 17.1 summarizes this situation by showing the ✓ symbols in parentheses to indicate that the criteria are fulfilled in a limited manner for all the universals. This finding suggests that the statistical universals that we examined in Parts II and III reveal two essential weak points of neural models: *rarity* and *long memory*.

---

[6]A possible direction to tackle this problem lies in the concept of subwords, in which a word is represented as a combination of a limited number of strings. There are interesting new works on the subword concept (Bojanowski et al., 2017).

[7]For the PTB, the quality of long memory is even weaker than for a standard literary text, because every article in the *Wall Street Journal* is short, consisting of at most several dozen sentences. Nevertheless, for consistency of data use in this chapter (especially with respect to Sect. 17.4), this section shows the PTB results. Even then, the three graphs for long memory in Fig. 17.3 show clear differences from those in Figs. 17.1 and 17.2. Takahashi and Tanaka-Ishii (2019) showed long-memory acquisition results from training with a text with stronger long memory. The qualitative message there was in common with the results in this chapter.

17.6 Future Directions for Generative Models

The last two chapters have described the quest for a generative mechanism that reproduces all the statistical universals. They have covered representative models that have been applied to reconsider language. After verifying every model of language, the primary understanding we have gained is that it is not at all easy to find a model that reproduces all the statistical universals considered in this book.

Nevertheless, we have seen that two generative models roughly reproduce all the statistical universals. These models are sophisticated; their mechanisms are based on complex systems.

Section 16.4 examined the generative model based on a random walk on a complex system (Tanaka-Ishii and Kobayashi, 2018). It introduces new instances naturally through visits to new nodes. The model enables natural extension of the network structure during a walk, and the scaling properties of population reflect the network structure. Furthermore, long memory arises naturally from the random walk, because it allows revisits to a node; nearby nodes have a strong chance of being revisited, causing clustering behavior.

Not just any network allows a random walk to reflect the statistical universals, however. The criteria are rather specific, and this is further evidence of how difficult it is to find a generative mechanism that reproduces all the statistical universals. In addition, a random walk cannot easily be applied in language engineering, because it does not have a learning procedure for constructing a model from a text. The question of how we can form a language model from a random walk on a certain network will require further study.

On the other hand, this chapter has shown the outcomes and consequences of past language models and ended by showing the potential of neural language models. For the first time in the history of language models, we have a language model that roughly covers almost all the statistical universals of language. Framed against this success, however, the previous section revealed the limitations of these models. In other words, the statistical universals highlight what neural language models are not good at. Specifically, they have only a limited capability to reproduce a complete population that follows the scaling properties, and they only roughly reproduce long memory.

Neural language models have mainly been studied in engineering, whereas the random walk on a complex network is in the realm of mathematics and physics. The frontiers of these fields, however, meet in numerous places. Both models have a complex network structure, and their advantages and disadvantages complement each other. The problems of one model have the potential to be tackled with the understanding of the other, and such study could lead to a neural network that would capture the self-similarities underlying language and enable the network to reproduce the statistical universals.

Part VI
Ending Remarks

Chapter 18
Conclusion

To question *what kind of system language is*, is tricky, because we are obliged to think in language. That is, we have to contemplate *what language is by using language*, or from a short vantage point, *what language is while being kept inside a language system*. Because we are thus enclosed in the system of language, our analysis method ultimately operates via language, but studying a method with the same method has some limits. If we only consider a target Z through the use of Z, we can never fully objectify or externalize it. One way to break this strong link between the method and its target is to *externalize* the target, and for this, we can observe aspects of the target that are beyond human control.

Currently, there are at least two approaches to externalizing the study of language. The first is neuroscience, which analyzes the cerebral signals resulting from stimuli. Because neural signals in the brain are phenomena related to language but are not directly controlled by a speaker, we can consider them to be external to the language system, and thus they facilitate an objective analysis of it. The other approach is to look for statistical universals, because they are present in all languages, and we are unaware that we produce them. Because the cerebral signals and statistical universals are outcomes of language use, it is reasonable to expect that their study could clarify the nature of language.

In general, a universal is defined as a property that holds for all natural language. Among such universals, this book has considered *statistical universals*, which are statistical properties revealed in large amounts of language data by following certain computational procedures. Such studies became possible with the capability to collect quantities of large-scale data and use powerful computers. Reflecting this macroscopic perspective on language, the book has taken a holistic approach by starting from a corpus, unlike a book on linguistics, where the typical approach is constructive, starting from phonemes and proceeding to words and then to sentences.

Statistical universals hold for various linguistic outputs of any language—at all times, by any author, and in any age. They even go beyond language and are found

K. Tanaka-Ishii, *Statistical Universals of Language*, Mathematics in Mind,
https://doi.org/10.1007/978-3-030-59377-3_18

in other human activities that use signs, such as music and computer programming. The book first overviewed these statistical universals. It then considered how they contribute to our understanding of language through the formation of linguistic units and phrase structures. Finally, it considered the nature of the statistical universals by studying various generative models of language. Through their examination, we have contemplated the singular properties of language. Such study is not only of scientific interest, but statistical universals can also be used to evaluate computational models for the purpose of language engineering. The statistical universals thus reveal the weak points of current engineering models.

Part I of this book began by situating the study of statistical universals within related fields and stating the aim of gaining a universal understanding about language by using computers and the methods of statistical mechanics. This study involves the academic fields of linguistics, computing, and complex systems theory. Chapter 2 compared statistical universals in relation to the general concept of *universals* in the field of linguistics, while Chap. 3 characterized language as a kind of complex system, amenable to the methods of statistical mechanics. Because language forms sequences of elements, the statistical universals naturally can be thought of as properties of (1) the vocabulary population and (2) the sequence itself. Chapter 3 also introduced two random sequences, a word-shuffled text and a monkey text, for the purpose of comparing their statistical properties with those of natural language.

Part II studied the vocabulary population of sets of words. Chapter 4 introduced the famous Zipf's law, i.e., the fact that the rank-frequency distribution follows a power law with a slope close to -1. Zipf's law implies that rare words have the same abundance no matter how large a text is. However, because Zipf's law almost holds even for a monkey text, we examined other properties that are absent from such monkey texts. These included properties of the rank-frequency distribution underlying consecutive sequences of length n, called n-grams. Chapter 5 more thoroughly considered the universality of these properties and biases in relation to Zipf's law, by examining over a thousand texts. The results verified that while the law is universal, there are deviations from it at least for large texts, child utterances, and character sequences.

After detailed analysis of the rank-frequency distribution, Chap. 6 then introduced two related power properties: the frequency distribution and the vocabulary growth (i.e., Heaps' law). It also explained their mathematical relation to Zipf's law. The phenomena here presented some deviations from the statistical mechanics underlying simple random texts. The chapter also summarized the state-of-the-art frontier of modeling the vocabulary population.

The statistical universals of the vocabulary population cannot distinguish a word-shuffled text from a natural language text. Because language forms a sequence,the way of distinguishing these texts should derive from the properties of the sequence. Part III therefore examined the properties underlying the sequence of elements such as words and characters. It introduced recent findings about language with respect to long memory, a notion of complex systems theory that indicates how an element influences another part of text a long distance away. Compared with the long history of population studies since Pareto, the study of long memory began only recently with the increased availability of large-scale data. Part III introduced three analysis methods, based on return intervals, long-range correlation, and fluctuation. All three showed that natural language texts behave differently from randomized texts created by shuffling the words of natural language texts.

In a natural language text, words have a larger probability of reuse, even after a very long distance. The reuse probability in a random text decays exponentially with the distance, but in a natural language text, this probability remains at a certain value. This fact implies that even rare words are not easily forgotten and always have a certain possibility of reuse. Chapter 7 examined this property in terms of the returns of words.

Natural language sequences also exhibit clusters, meaning that words occur abundantly in some parts of a text but scarcely in others. One factor in such clustering phenomena lies in the flow of context. The long-range correlation described in Chap. 8 and the fluctuation analysis methods of Chap. 9 can capture the degree of clustering.

The long-range correlation analysis explained in Chap. 8 is a technique to quantify the length of memory. Precisely, it examines the correlation decay between two subsequences with respect to the distance between them. If the decay follows a power function, then the phenomenon is called long memory, and it indicates how two subsequences may be similar even when they are a long distance apart. This phenomenon has frequently been analyzed for numerical time series that are not natural language, by using the autocorrelation function. As natural language is nonnumerical, this chapter applied the autocorrelation function to the return interval sequence. The analysis presented a power decay, thus indicating that long memory universally exists in natural language texts. The overall evidence further showed that the long-range correlation of natural language is weaker than that in appearing in other real-world complex sequences formed of a small, finite number of elements.

On the other hand, fluctuation analysis considers measuring the degree of variance in comparison with independent and identically distributed (i.i.d.) sequences. The variance becomes comparatively larger when there is a clustering phenomenon. Chapter 9 thus overviewed various methods to quantify this degree, which all presented a power law. Because the power law holds even for an i.i.d. process, the clustering phenomena of a natural language sequence can be considered to derive from amplification of the natural fluctuations that already exist in the underlying probabilistic process. The chapter empirically showed that the power exponent quantified through fluctuation analysis can distinguish a real text from an i.i.d. process. It can even distinguish different kinds of real texts, such as writing and speech. The exponent can be metaphorically interpreted as a kind of fractal dimension. The overall degree of fluctuation of a written text in a human language, however, was shown not to be emphasized in comparison with that of an i.i.d. process; i.e., it is weak, complementing the understanding gained from the previous chapter.

Finally, at the end of Part III, Chap. 10 introduced the Shannon entropy rate to represent all the characteristics explained in Parts II and III. The entropy rate shows the predictability of the element succeeding a sequence. It decreases in accordance with a power function with respect to the text length, and the exponent indicates a degree of difficulty for learnability and shows similar values across languages; thus, it is possibly another statistical universal. The entropy rate turns out to be an aggregate property, however, of all the statistical universals considered thus far. Furthermore, it shows the quality of the language model used to acquire it, in addition to the quality of the language sequence. Therefore, a study of language sequences highlighting the specific characteristics underlying natural language requires a different approach from this notion of the entropy rate, which was attempted in the first three chapters of Part III.

Next, Part IV explained how the statistical universals considered thus far would contribute to the formation of linguistic elements such as words and phrase structures. It also introduced two linguistic universals found by structural linguists and their use in the field of computational studies. Their relations with the statistical universals were also discussed.

Chapter 11 discussed how words could arise from the statistical universals. Harris (1955) found that word boundaries correlate well with statistical information that has little relation with meaning. While Chap. 10 showed a universal decrease in the complexity of the next character with respect to the context length, Harris's hypothesis implies that locations that violate this universal decrease correlate with the boundaries of linguistic units such as words and morphemes. The findings of a large-scale empirical verification conducted in the field of computational linguistics in an effort to reformulate Harris's scheme supported his hypothesis, and they implied that language innately possesses a structure for articulation that is related to a statistical universal underlying the complexity of language.

Chapter 12 considered how statistical universals might contribute to the meanings and values of words. One way to formulate values for words is via the distributional semantics, again proposed by Harris, which defines the meaning of a linguistic unit by its distribution. The chapter considered two values of linguistic units along this line. First, it considered using the frequency of a word as its value. Here, the Weber–Fechner law is known to relate a stimulus to its perception, and the concept was applied to contemplate the effect of frequency in language. The logarithm of the frequency of words was shown to correlate well with their familiarity, a human perception about a word, although there were limitations due partly to the finite size of the corpus.

Second, Chap. 12 considered the distributional hypothesis, a theory in distributional semantics, which suggests that *similarly distributed words have similar meanings*. Following a recent trend of representing a word by its context word vector, Tian et al. (2017) mathematically deduced an upper bound on the compositionality of meaning in terms of the vectors of two neighboring words and their sum, for the case when Zipf's law holds among elements. Their formulation shows how the distance between two word vectors becomes small when they appear more exclusively as neighbors. This shows how Zipf's law provides the basis of the compositionality of meaning; however, the question of whether the compositionality truly holds depended on the way words occur in the clustering phenomena described in Part III.

Chapter 13 examined the relation between the frequency and size of a word. Previously, Zipf showed that infrequent words are longer and used this fact to reason how language is economical. However, the phenomenon that infrequent words are longer is trivial in a simple random sequence, such as a monkey text. Chapter 13 examined the obvious fact that linguistic units are much longer than those appearing in simple random sequences because of the inherent compositional nesting of linguistic units.

Chapter 14 considered the relation between the statistical universal of long memory and phrase structure. It introduced the theory of Lin and Tegmark (2017), who analytically proved that a simple grammatical structure can induce long memory, depending on the grammar's setting. The empirical evidence examined in Part III, however, implied a weaker long memory underlying language sequences than what their proof showed. The Greenberg universal for word order, a linguistic universal about sentence structure, functions to suppress long memory within a sentence, which could partly explain why long memory is weak. This universal only applies to sentences, however, whereas natural language exhibits weak long memory even at a scale of thousands of words. Language should thus have a more expansive structure across sentences that produces long memory. Hence, the last part of the book contemplated that mechanism via generative models.

Part V overviewed possible causes of the statistical universals. Chapter 15 considered theories that explain Zipf's law, the oldest of the statistical universals. In particular, since Mandelbrot (1953), an argument has existed that Zipf's law results from some optimization for communication. Although such a mathematical interpretation is interesting, it does not necessarily indicate that language production is the outcome of optimization. Furthermore, such optimization would require an optimal word choice even in the middle of discourse, which seems unlikely for some cases such as infant utterances.

Another, more convincing approach to the statistical universals underlying population is, rather, to consider that they hold because of a statistical consequence, specifically, the consequence of some mathematical limit theorem. Mandelbrot (1997) indicated that power laws are invariant across various mathematical transformations. In other words, once we sample from a power-distributed vocabulary, the resulting phenomena become power distributed. If the statistical universals hold because of such a limit theory, then they are precursors of language and also contribute to formulating language. Part IV discussed such possibilities. Furthermore, the singular aspects of sequences that otherwise share the same statistical properties are important. As Stumpf and Porter (2012) indicated, one way to investigate these singular aspects is through generative models of language.

Chapters 16 and 17 therefore examined representative generative models, namely, mathematical and language models. The goal of these chapters was to contribute to language engineering by using the statistical universals to evaluate the limitations of representative models to date. Both chapters showed the great difficulty of finding a generative process that accounts for all the statistical universals examined in this book. Among the models examined in this book, only two processes based on complex systems show the potential to produce all the universals: random walks on a specific complex network and state-of-the-art neural models. As both are based on large-scale complex mechanisms, these findings suggest that our language system might require a generative process based on a large-scale complex model. Although both processes have limitations in reproducing all the statistical universals, the pros and cons of each complement each other. Hence, Part V ended by proposing a direction for future studies, to find a generative language model that can reproduce all the statistical universals well. Moreover, because language models are the basis of computational linguistics, the statistical universals could be used as a means to evaluate language models in language engineering.

Overall, I should emphasize that statistical universals are partly a mathematical consequence of statistical mechanics. We have seen how researchers in the past may have viewed statistical universals as revealing some mysterious aspect attributed to language. This book, however, is meant to challenge that way of thinking. Rather, the statistical mechanics functions as a precursor of language, and such a mechanism thus accommodates language. Then, the study of language, a system exclusive to humans, must clarify how that precursor stipulates or formulates language and how language deviates from trivial statistical consequences. In other words, the study of statistical universals reveals how language depends on a statistical precursor. Hence, one path for future studies is to generate random sequences that reproduce all the statistical universals. By understanding how different natural language is from these random sequences and contemplating how this difference is essential to the formulation and perception of language, we could gain a deeper understanding of the characteristics of human language.

Given the findings presented in this book, we might suppose that language is similar to many other natural and social systems. The statistical universals we have examined here have been reported to underlie phenomena studied in the natural and social sciences. Language is thus not so unique compared with other natural and social systems. Rather, the particularity of language is that its main form is the sign. A sign signifies, and therefore, it directly describes an external system, making it tractable via language. Language thus assimilates various other systems. The limitation shown in this book, however, is that language has trouble assimilating itself. As stated at the beginning of this chapter, the reason is that the descriptive framework and the target become the same for the case of language. Here, statistical universals could provide a way to bypass this limitation. I believe that the study of language universals and generative models, from both scientific and engineering perspectives, points to a new horizon in understanding and processing language.

Chapter 19
Acknowledgments

The motivation behind this book appeared long ago in my adolescence, when I first encountered the question of how language forms a fractal. Being a student of applied mathematics and at the same time interested in language, I wondered how language could be self-similar. It has taken me all these years to give form to this question and its potential answers. I could never have continued studying this question for so long without the support and kind hearts of the people mentioned here.

First, I thank Prof. Shigeo Kusuoka, a mathematician and emeritus professor of The University of Tokyo, for our academic discussions. Meetings with him have been among the most thoughtful moments of my life. He also guided me to consider generative models, which culminated in the work presented in Part V. I also thank Prof. Armin Bunde, Professor of Physics at the University of Giessen, who introduced me to the interval analysis presented in Chap. 8. It was essential training for me during 2015–2016 on how to conduct research in statistical mechanics.

Among my colleagues, I thank Prof. Łukasz Dębowski of IPIPAN (Polish Institute of Computer Science of the Polish Academy of Sciences), who kindly reviewed my book thoroughly and gave me essential comments indeed. I also express gratitude to Prof. Satoru Iwata of the Graduate School of Information Science and Technology at The University of Tokyo, who went through the manuscript at a very early stage and gave me fruitful suggestions. Furthermore, I thank Prof. Kyo Kageura of the Graduate School of Education at The University of Tokyo for his valuable philosophical comments and discussions. I was also extremely lucky to have the comments of Prof. Takashi Tsuchiya of GRIPS (National Graduate Institute for Policy Studies), who kindly checked my manuscript from a mathematical perspective. His comments were indeed very valuable and helped greatly to improve the content of this book.

My gratitude is also due to Mr. James See, who helped me to improve the text quality. It has been amazing to have someone go through all my diverse thoughts for over 15 years. The articles that are the basis of this book were finished with his

© The Author(s) 2021
K. Tanaka-Ishii, *Statistical Universals of Language*, Mathematics in Mind,
https://doi.org/10.1007/978-3-030-59377-3_19

help, and I feel grateful for his strong support. I also thank Mr. Steve Armstrong and Ms. Tomoko Nishigaki for helping me to improve the quality of the text.

I also express my sincere gratitude to Ms. Yasue Okajima, who cared for me so that I could have time to work on this book. I thank my students, at Kyushu University and The University of Tokyo, with whom I worked. They worked on some of the studies mentioned in Chaps. 9–12, 16, and 17. Their work included not only conducting the actual studies but also assembling the data necessary for the research, and it involved a great deal of trial and error. In alphabetical order, the colleagues who collaborated with me on this project are as follows: Shunsuke Aihara, Daiki Hirano, Daisuke Kimura, Tatsuru Kobayashi, Hiroki Matsuura, Geng Ren, Shuntaro Takahashi, Ryosuke Takahira, Hiroshi Terada, and Jin Zhihui. It was, above all, an extraordinary experience to have such young colleagues with whom I could share and shape some common academic interests.

I was offered the opportunity to publish this book in the series *Mathematics in Mind* by Prof. Marcel Danesi of the University of Toronto. Prof. Danesi has kindly supported my work ever since the publication of my previous book. This book never would have been accomplished without his support. I thank Ms. Dahlia Fisch and Ms. Elizabeth Loew of Springer for editing the book. I also thank Mr. Jeffrey Taub, Mr. Herbert Moses, and Ms. Ramabrabha Selvaraj for typesetting and producing this book. Furthermore, this book has an extended Japanese version, and I thank Mr. Kensuke Goto for his support in realizing both volumes.

I would not have been able to author the book if I did not work for the Research Center of Advanced Science and Technology (RCAST) at The University of Tokyo. I thank the Japan Science and Technology Agency (JST) for the financial support to conduct basic research through the PRESTO and HITE projects (grant numbers JPMJPR14E5 and JPMJRX17H5). The book was also supported by the Suntory Foundation.

Lastly, I was most happy to have my family accompany me on this journey. I was often lost in a multidisciplinary forest, but the light they brought helped me to see the path to a clearing.

In sum, I express my deepest gratitude to all of these people for supporting me in this endeavor.

Part VII
Appendix

Chapter 20
Glossary and Notations

This appendix provides a glossary of the vocabulary and definitions of the notations used in this book. The definitions of linguistic terms often differ here from conventional notions. On the other hand, most of the mathematical notions follow convention.

The entries here mention the most related part or chapter, if applicable.

20.1 Glossary

Chance

Chance in this book denotes a probabilistic choice. When there are multiple options, one is chosen according to the probabilities for each item.

Complex system

A complex system is one for which it is difficult to understand the global behavior from only the behavior of its parts and elements. In a complex system, events at different scales interact (Thurner et al., 2018). One consequence is a set of power laws at a macroscopic scale. Accordingly, Bak et al. (1987, 1988) defined a complex system as a system that fulfills power laws. Following this definition, a complex system in this book means a system that has a scale-free structure in its element population and produces long memory (Chap. 3).

The original version of this chapter was revised. The correction to this chapter is available at https://doi.org/10.1007/978-3-030-59377-3_24

© The Author(s) 2021, corrected publication 2022
K. Tanaka-Ishii, *Statistical Universals of Language*, Mathematics in Mind,
https://doi.org/10.1007/978-3-030-59377-3_20

Conditional probability

The conditional property of X given Y, denoted as $P(X|Y)$, signifies the probability that X occurs under the condition that Y is given. Mathematically, $P(X|Y) \equiv P(X, Y)/P(Y)$. See the entry on **probability** too (Chap. 10).

Corpus

A corpus is a text or a set of texts. Representative corpora are a collection of newspaper articles and an individual text such as *Moby Dick*. Usually, a corpus is assembled manually so that the texts in it meet certain criteria. This book considers a corpus as a sequence of words or characters, as defined in Chap. 3.

Distribution

A distribution in this book describes the numbers of possible events that should occur when effectuating a large number of trials. A distribution is often described by a function. For example, the numbers of events for the six faces of an ideal die should be almost the same across the faces. A uniform function describes this distribution.

The concept presumes a true underlying source, but in reality only samples are acquired as data. As the true source is usually unknown, a distribution is acquired through assumptions based on some prior knowledge or through data.

The main distributions that appear in this book are defined by power and exponential functions, introduced in Chap. 3 and Chap. 7, respectively.

Fluctuation

Fluctuation is the change of a phenomenon over time or space. In this book, fluctuations indicate changes in word occurrences, meaning that in some parts of a text, a word or a set of words occurs abundantly, whereas in other parts, it hardly occurs (Chap. 9).

Function

A function is a mapping of an element x of a set X (the domain) to an element y of another set Y (the range). For example, a rank-frequency distribution is a function mapping a rank r (the domain) to its frequency f, with the functional relation $f \propto r^{-\eta}$ (Chap. 3).

Generative model

A generative model is a mathematical model that generates a sequence of elements. It usually defines a subsequent element, conditioned on the past sequence. The simplest generative model in this book is the model to produce a monkey text (Sect. 3.5), which produces each character independently of the previous characters, following some character distribution. As for more complex generative models, Chap. 16 introduces mathematical models of language, while Chap. 17 introduces language models, which are probabilistic generative models that produce a word sequence.

Law

A law in this book indicates a stylized hypothesis empirically acquired through observation. The term appears in the context of a *power law*, or a *scaling law*, mainly used in the field of statistical mechanics, which is a distribution following a power function. In physics, the use of the term *law* presumes an underlying theory. Unfortunately, we do not yet fully understand why power laws of language hold; therefore, the true signification of the term *law* in this book is closer to that of a stylized hypothesis, i.e., a common observation without a solid theoretical foundation (Sect. 2.3).

Least-squares method

This is a method to estimate functional parameters by minimizing the square error (Sect. 21.1).

Long memory

Long memory is a phenomenon of sequences in which one occurrence in a sequence influences another part of the sequence at a long distance. Long memory has various causes, as mentioned in Sect. 3.4. In this book, long memory is mainly considered in terms of clustered occurrences of words. Changes in context are the most important cause of long memory (Sect. 3.4, Part III).

Long-range correlation

Long-range correlation is a long-memory phenomenon in which two subsequences of a sequence at a distance s are correlated with only a power decay with respect to s. This implies that the two subsequences, even if they are separated at a long distance, are correlated. For a simple stochastic process, such as a Markov process, two parts of a sequence are usually only correlated for a small s, and the degree of correlation decays exponentially with respect to s (Sect. 3.4, Chap. 8).

Long-range dependence   This book uses long-range dependence to indicate
                             that a word modifies another word at a distance in
                             the same sentence (Sect. 3.4, Chap. 14).

Maximum-likelihood method   This is a method to estimate functional parameters
                             by maximizing the likelihood (Sect. 21.1).

Meaning   Meaning is the value of some linguistic unit such
                             as a word, sentence, or text. It is a difficult task to
                             define what exactly this value is. Parts II and III of
                             this book conduct an analysis devoid of meaning
                             in order to reveal statistical universals. Part IV,
                             especially Chap. 12, reconsiders what meaning
                             could be by investigating how statistical universals
                             contribute to the formulation of meaning.

Model   A model is a mathematical description of a phe-
                             nomenon. It is no more than a hypothesis acquired
                             by simplifying and approximating observations of
                             the phenomenon. The book considers models of
                             language, of which there are two kinds: generative
                             models, which describe how language is produced,
                             as discussed in Part V, and mathematical mod-
                             els based on statistical universals. Zipf's law, for
                             example, is a model of the rank-frequency distribu-
                             tion.

Natural language   Natural language means language that humans use
                             for communication. The modifier *natural* is applied
                             when necessary for contrast with artificial lan-
                             guages such as programming languages.

Power function   A power function is a specific kind of function
                             defined in Sect. 3.2, and it is the main function
                             considered in this book. Power functions are used
                             in models of statistical universals.

Probability   The likelihood of an event w occurring, among a
                             set of possible events, W, is called the probability
                             and denoted as $P(w)$. $P(w)$ is subject to conditions
                             such as $0 \leq P(w) \leq 1$ and $\sum_{w \in W} P(w) = 1$.
                             This book considers language to be a sequence
                             $X = X_1, \ldots X_i, \ldots X_m$, as defined in Chap. 3,
                             where X_i denotes the ith element of X. The prob-
                             ability of X_i being $w \in W$ is denoted as $P(X_i = w)$.

Process

A process generates a sequence. This book considers a text as an outcome of a linguistic process. When the generative process is probabilistic—that is, items are chosen by chance following a probability distribution—the process is called a *probabilistic process* or *stochastic process*.

Random

Randomness in this book means the selection of an element by chance from a set of possible elements according to some probability (Chap. 3).

Return

A return is the reappearance of a particular event in a sequence. The number of other events occurring before the return defines a return interval length (Chap. 7).

Scale free

This term, *scaling*, and *scale invariant* describe a property of a system that holds independently of the system's size. They both imply that the property is invariant with respect to enlargement or reduction of the system size. The concept is synonymous with the term *self-similar*, but the scale-free and scale-invariant notions highlight the invariability with respect to the system size.

Self-similarity

Self-similarity signifies a property of a system in which the whole includes itself as a part. The Koch curve in Fig. 3.2 gives one example. It is infinitely self-similar, meaning that the curve includes its own shape an infinite number of times at different scales. On the other hand, natural and social systems may have a self-similar property, but it is often finite. The property can become observable through statistical transformation of a phenomenon, in which case the term *statistically self-similar* is more appropriate (Sect. 3.3).

Sequence

A sequence is a series of elements aligned in one-dimensional order. An element in this book is a linguistic unit, such as a word, a character, or a phoneme. When the elements are aligned over time, the sequence is called a time series. Note that this book does *not* consider language as a time series. A sequence is either continuous or discrete, and the latter category includes both numerical and nonnumerical types. Language is thus a discrete, nonnumerical sequence (Sect. 3.1.2).

Sentence

A sentence is a sequence consisting of multiple words. In language, a sentence is a unit that produces a unified meaning.

System

According to the Oxford Dictionary, a system is *a set of things working together as parts of a mechanism or an interconnecting network, a complex whole.* This book regards language as such a system.

Text

A text is a sequence of sentences, each of which consists of a sequence of words. A text typically has content.

Universal

A universal is a property that holds across any sample of the target in question. For the case of natural language, a language universal is a property of language that holds across all natural language samples. Chapter 2 gives some examples. This usage comes from the field of linguistics. In contrast, the topic of this book is statistical universals of language, a kind of universal that is only observable by performing statistical analyses of texts (Sect. 2.3).

Word

A word is an entity of language that typically is meaningful. Sentences are composed of consecutive sequences of words (Chap. 3).

20.2   Mathematical Notation

The default mathematical notations throughout this book are as follows:

Variables: Greek letters denote important functional parameters. Sets and probabilistic variables are denoted in uppercase. Other variables are denoted in lowercase by default and are used consistently throughout the book, except for the usages of a, b, g, i, j, k, t, x, and y. Variables are in italic face.

Functions   are mostly denoted in uppercase, except for variables that are treated as a function of another variable, and functions that are introduced provisionally and locally for explanatory purposes. Variable functions are given in *italics* by default; however, following mathematical convention, invariable functions such as the Shannon entropy, expectation, and variance, whose definitions are unique and common in mathematics, are presented in normal type.

Because of limitations in the application software for figure production, the text in some figures does not conform to the above conventions.

The following notations are used consistently throughout the book:

C Set of characters (or phonemes in Chap. 11), with size $|C|$. An individual character is denoted as c

f Frequency of a word or a character

h Entropy rate (Chap. 10)

l Window size (Chap. 9)

m Text length

n General indication of length, for n-grams, for a consecutive sequence of length n, or for a length of n preceding elements (Chaps. 4, 10, 11, 17)

q Return interval length (Chaps. 7, 8)

Q Sequence of interval lengths, denoted as $Q = Q_1, Q_2, \ldots, Q_i, \ldots, Q_{|Q|}$, where Q indicates the entire sequence and Q_i indicates the variable of the ith interval length in Q (Chaps. 7, 8)

r Rank of a word by its frequency (Part II, Chap. 13)

s Distance between two subsequences in long-range correlation analysis (Chap. 8)

u Alphabet size, equal to $|C|$

v Vocabulary size, equal to $|W|$

W Set of words, with size $|W|$. An individual word is denoted as w

X Sequence of words or characters, denoted as $X = X_1, X_2, \ldots, X_i, \ldots X_m$, where X indicates the entire sequence and X_i indicates the variable of the ith element in X

z Relative frequency of a word or a character

α Taylor exponent (Chap. 9)

β Hilberg ansatz decay constant (Chap. 10)

η Negative of the exponent of a rank-frequency distribution (Chaps. 4, 5)

γ Negative of the exponent of long-range correlation when measured by the autocorrelation function (Chap. 8)

κ Functional parameter for the power term of a stretched exponential function (Chap. 7)

λ Mean interval length, i.e., $\lambda \equiv \frac{m}{f}$ (Chap. 7)

μ Mean of some random variable (Chap. 9)

ν Exponent of fluctuation analysis by the Ebeling–Neiman (EN) method (Chap. 9)

ψ Rare word fraction applied to a sequence of length m (Chaps. 7, 8)

σ Standard deviation of some variable (Chap. 9)

θ Functional parameter for the exponential term of a stretched exponential function (Chap. 7)

ξ Exponent of vocabulary growth, i.e., Heaps' exponent (Chap. 6)

ζ Negative of the exponent of a density function (Chap. 6)

$\propto$ $y \propto x$ means that $y = ax$ for some constant a

$\approx$ $y \approx x$ means that y is approximately equal to x

$\equiv$ $y \equiv x$ means that y is defined as x

$\rightarrow$ $x \rightarrow a$ indicates the value of x approaching infinitely close to a

$\Rightarrow$ $\Rightarrow$ indicates mathematical convergence (Chaps. 10, 21)

$\rightarrowtail$ $a \rightarrowtail b$ indicates a grammatical rewriting rule (Chap. 14)

20.3   Other Conventions

Note that various figures in the text use color in specific ways, especially figures showing plots of statistical universals measured for a corpus. In those figures, plots of primary importance for a natural language text appear in red. Other plots use other colors to highlight differences. In particular, plots for random sequences (e.g., shuffled and monkey texts) are always in yellow. Chapters 16 and 17 use summary figures to reconsider the results of primary importance, but the plots are in yellow because they were calculated for random sequences. In any figure, a black line represents an analytical line, whereas a thick gray line indicates a fitting line.

In this book, unless mentioned otherwise, the natural logarithmic base is used. Estimated values are reported to three significant figures after the decimal point for direct results appearing in graphs and footnotes with occasional exceptions where necessary. When the text mentions such values, or for results subjected to a secondary analysis, only two significant figures after the decimal point are shown.

Chapter 21
Mathematical Details

This appendix summarizes the mathematical details of the arguments presented in the main text. As this summary is brief, readers who would like a more thorough presentation should refer to the corresponding textbooks and original papers cited below.

21.1 Fitting Functions

Acquiring a good fit function is one way to investigate the properties underlying a set of data points. The following two methods are the most basic ones that can be used to estimate functional parameters, given some functional form.

The least-squares method estimates functional parameters by minimizing the square error. Let x denote an input and y denote an output. For k observations $(x_1, y_1),\ldots, (x_k, y_k)$, suppose that a functional relation $y \approx g(\Psi, x)$ is hypothesized, where Ψ denotes the parameters of the function g. The least-squares method then estimates the following:

$$\hat{\Psi} = \arg\min_{\Psi} \varepsilon(\Psi),$$

$$\varepsilon(\Psi) \equiv \sqrt{\frac{1}{k}\sum_{i=1}^{k}(y_i - g(\Psi, x_i))^2} \; .$$

When g is a power function, Clauset et al. (2009) indicated problems in fitting with the least-squares method. Consequently, this book instead uses the maximum-likelihood method (see below) when y can be defined as a probability. Otherwise, to fit a power function with the least-squares method, the right side of the second formula takes a logarithmic form:

© The Author(s) 2021
K. Tanaka-Ishii, *Statistical Universals of Language*, Mathematics in Mind,
https://doi.org/10.1007/978-3-030-59377-3_21

$$\varepsilon(\Psi) \equiv \sqrt{\frac{1}{k} \sum_{i=1}^{k} (\log y_i - \log g(\Psi, x_i))^2}.$$

As the square error ε, called the residual, indicates an observation's goodness of fit by the least-squares method, the main text of the book reports ε for the best set of parameters.[1]

The maximum-likelihood method estimates functional parameters by maximizing the likelihood. For a set of k data $x_1, \ldots, x_k$, where each probability is defined by a probability function $P(x_i|\Psi)$, the maximum-likelihood method estimates Ψ as follows:

$$\hat{\Psi} = \arg \max_{\Psi} L,$$

$$L \equiv \prod_{i=1}^{k} P(x_i|\Psi).$$

L is called the *likelihood* and is one indicator of the goodness of fit for this method. The negative logarithm of the likelihood is usually used instead, however, as follows:

$$LL \equiv -\log L. \tag{21.1}$$

A smaller LL means a better fit. This book reports the minimum value of LL (i.e., when L is at a maximum), for the best set of parameters, whenever it describes an application of the maximum-likelihood method.

21.2   Proof that Monkey Typing Follows a Power Law

The related part in the main text appears in Sect. 4.3, page 38.

Miller (1957) considered a monkey that randomly types any of u characters, which consist of a space and multiple non-space characters. Because the space separates words, let its probability be a, and then let each of the other characters be hit uniformly with a probability of $(1 - a)/(u - 1)$. The number of words of length k is $(u - 1)^k$. Although there is no guarantee that all these possible words occur, the rank r_k of a word of length k is roughly constrained by the following inequality, because longer words are less likely to occur:

---

[1] Note that when g is an exponential function, as in Chap. 7, it is fitted by taking the logarithm only for y.

$$S(k) + 1 \leq r_k < S(k+1), \tag{21.2}$$

where

$$S(k) \equiv \sum_{i=1}^{k-1} (u-1)^i = \frac{(u-1)^k - (u-1)}{u-2}. \tag{21.3}$$

Therefore, the rank r_k of a word of length k grows roughly exponentially with respect to k, i.e.,

$$r_k \approx (u-1)^k. \tag{21.4}$$

Given that the probability of occurrence of a word of length k is $a \left(\dfrac{1-a}{u-1} \right)^k$, by replacing k with the rank r_k via the relation given in formula (21.4), we obtain the rank-probability distribution as

$$P(r_k) = a \left(\frac{1-a}{u-1} \right)^{\log r_k} = a r_k^{\log(1-a)-1}, \tag{21.5}$$

where the log is taken with base $u - 1$. This result shows that the probability distribution follows a power function with respect to the rank.

21.3   Relation Between η and ζ

The related part in the main text appears in Sect. 6.1, page 56. This section provides a proof of formula (6.2).

Lü et al. (2010) deduced the relation between the exponents of the rank-frequency distribution and the density function, as follows. First, Fig. 21.1 shows a graph of a rank-frequency distribution. By counting the number of points in the rectangle in Fig. 21.1 from the two perspectives of r and f, we obtain the following equation:

$$\delta r = P(f(r))\delta f. \tag{21.6}$$

Because $f \propto r^{-\eta}$ from formula (4.1),

$$\delta f \propto r^{-\eta} - (r + \delta r)^{-\eta} \approx r^{-\eta-1}\delta r.$$

Fig. 21.1 Schematic
explanation of formula (21.6)

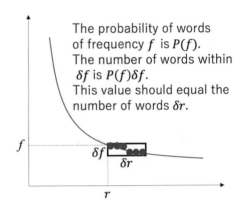

The probability of words
of frequency f is $P(f)$.
The number of words within
δf is $P(f)\delta f$.
This value should equal the
number of words δr.

Therefore,

$$P(f) = \frac{\delta r}{\delta f} \propto r^{\eta+1} = (r^{-\eta})^{-\frac{\eta+1}{\eta}} \propto f^{-\frac{\eta+1}{\eta}} = f^{-\zeta},$$

and thus,

$$\zeta = 1 + \frac{1}{\eta}. \qquad (21.7)$$

21.4   Relation Between η and ξ

The related part in the main text appears in Sect. 6.2, page 60. This section provides
a proof of formula (6.5), again following the deduction by Lü et al. (2010).

As shown in Fig. 6.2, the total number of words that have frequencies between
f_{max} and f, where $f \leq f_{max}$, almost equals the rank of the word whose frequency
is f. In other words, because the shaded area under the curve equals the rank, we
have

$$r \approx \int_{f}^{f_{max}} vP(f)df = av[f^{1-\zeta} - f_{max}^{1-\zeta}], \qquad (21.8)$$

with a indicating some constant. Given that $\zeta = 1 + \frac{1}{\eta}$,

$$f^{1-\zeta} - f_{max}^{1-\zeta} = (f_{max}r^{-\eta})^{1-\zeta} - f_{max}^{1-\zeta} = f_{max}^{-\frac{1}{\eta}}(r - 1). \qquad (21.9)$$

From (21.8), we can deduce the following relation:

$$f_{max} \propto v^{\eta}. \qquad (21.10)$$

Given that integration over all frequencies would equal the text length m, and that $f = f_{max} r^{-\eta}$,

$$m = \sum_{r=1}^{v} f(r) \approx \int_{1}^{v} f(r) dr = \frac{f_{max}(v^{1-\eta} - 1)}{1 - \eta}. \tag{21.11}$$

Then, we have the following equation:

$$\frac{v^{\eta}(v^{1-\eta} - 1)}{1 - \eta} \propto m. \tag{21.12}$$

From this relation, let us next investigate the relation of η with ξ. Recall Heaps' law (6.4):

$$v \propto m^{\xi}. \tag{21.13}$$

By using this relation, formula (21.12) can be roughly solved as separate cases for $\eta \gg 1$ and $\eta \ll 1$. For the case of $\eta \gg 1$,

$$v \propto ((\eta - 1)m)^{1/\eta}, \tag{21.14}$$

so

$$\xi \approx 1/\eta. \tag{21.15}$$

For the case of $\eta \ll 1$, $v \propto (1 - \eta)m$, meaning that

$$\xi \approx 1. \tag{21.16}$$

21.5 Proof That Interval Lengths of I.I.D. Process Follow Exponential Distribution

The related part in the main text appears in Sect. 7.4, page 70.

Consider a shuffled text in which there is no dependence among words, meaning that words occur independently. Given the frequency f of a word and the text length m, let the mean length be $\lambda \equiv \frac{m}{f}$. The probability of a word occurring is then $\frac{1}{\lambda}$. The probability of a word having interval length q is

$$P(q) = \frac{1}{\lambda}\left(1 - \frac{1}{\lambda}\right)^{q} \approx \frac{1}{\lambda}\exp(-q/\lambda). \tag{21.17}$$

The last transformation gives a known case for sufficiently large λ. The resulting distribution is known as an exponential distribution.

For a sequence whose elements occur following an i.i.d. process, therefore, the interval length of an element follows an exponential distribution. This implies that the number of times that an element type occurs in a segment follows a Poisson distribution. This is an important premise for the theoretical considerations in Chap. 9.

21.6   Proof of $\alpha = 0.5$ and $v = 1.0$ for I.I.D. Process

The related parts in the main text appear in Sect. 9.1, page 90, and Sect. 9.2, page 91.

Consider i.i.d. variables $X_1, \ldots, X_i, \ldots, X_m$, where i denotes the location within a text. For a specific word $w_k \in W$, with W being the set of words, let p_k denote the probability of occurrence of w_k, i.e., $P(X_i = w_k) = p_k$ (for all i). Naturally, the expectation E and variance Var of the count of w_k for X_i are

$$E[X_i = w_k] = p_k, \tag{21.18}$$

$$\text{Var}[X_i = w_k] = p_k(1 - p_k), \tag{21.19}$$

which only depend on the constant p_k. Next, let $y(w_k, l)$ denote the count of w_k occurring within a window of size l, as defined in the main text. Obviously, then, $E[y(w_k, l)] = l E[X_i = w_k]$. The variance of the word w_k occurring within the window of size l is as follows:

$$\text{Var}[y(w_k, l)] = \text{Var}\left[\sum_{i=1}^{l} X_i\right]$$

$$= E\left[\left(\sum_{i=1}^{l}(X_i - p_k)\right)^2\right]$$

$$= E\left[\sum_{i=1}^{l}(X_i - p_k)^2 + 2\sum_{i \neq j}(X_i - p_k)(X_j - p_k)\right]$$

$$= E\left[\sum_{i=1}^{l}(X_i - p_k)^2\right]$$

$$= \sum_{i=1}^{l} \text{Var}[X_i = w_k]$$

$$= l\text{Var}[X_i = w_k].$$

Note that for the third to fourth lines above, $E[(X_i - p_k)(X_j - p_k)] = 0$ for every i, j with $i \neq j$, because X_i and X_j are independent of each other and formula (21.18) follows. Therefore,

$$\text{Var}[y(w_k, l)] = \frac{\text{Var}[X_i = w_k]}{E[X_i = w_k]} E[y(w_k, l)]. \tag{21.20}$$

Because both $E[X_i = w_k]$ and $\text{Var}[X_i = w_k]$ depend only on the constant p_k, the Taylor exponent α of an i.i.d. process is 0.5.

Furthermore, for the EN exponent, $y(l)$ is the sum of the $\text{Var}[y(w_k, l)]$ for different words w_k. Therefore,

$$y(l) \equiv \sum_{w_k \in W} \text{Var}[y(w_k, l)] \tag{21.21}$$

$$= l \sum_{w_k \in W} \text{Var}[X_i = w_k]. \tag{21.22}$$

The summed term does not depend on l. Hence, the EN exponent ν of an i.i.d. process is 1.

21.7 Summary of Shannon's Method to Estimate Entropy Rate

The related part in the main text appears in Sect. 10.2, page 103. The base of the logarithm in this section is 2.

When a subject guesses the character following a given string of length n, the answer will be correct or incorrect. The subject thus obtains a *prediction* of X_n by making multiple guesses, one character at a time, until he/she reaches the correct answer. In other words, a prediction for a character X_n in this setting consists of a series of guesses.

The number of guesses required to reach the correct answer reflects the predictability of that character; specifically, it should relate to the probability of character X_n appearing after X_1^{n-1}. Let k_i^n denote the probability that a subject requires i *guesses* in a prediction to find the correct letter following a block of length $n - 1$.

For this setting, Shannon deduced the following inequality (Shannon, 1951):

$$\sum_{i=1}^{u} i(k_i^n - k_{i+1}^n) \log i \leq H(X_n | X_1^{n-1}) \leq - \sum_{i=1}^{u} k_i^n \log k_i^n. \tag{21.23}$$

Here, u is the number of characters in the set; in his work, $u = 27$, because the English alphabet consists of 26 letters and a space. Note that this lower bound is actually the lower bound of the upper bound of $H(X_n|X_1^{n-1})$ and not the direct lower bound. For each context length n, the probability k_i^n can be calculated for a set of samples.

In Shannon's original experiment, he took 100 phrases of length 100 from *Jefferson the Virginian*, a biography of ex-US President Thomas Jefferson authored by Dumas Malone. In each experimental session, he asked a subject (his wife, according to Moradi et al. (1998)) to predict the next character, given a block of length $n - 1$. She continued in this manner for $n = 1, 2, \ldots, 15$, and 100 for each phrase; consequently, Shannon acquired 16 observations for each phrase. He then calculated k_i^n for $n = 1, 2, \ldots, 15$, and 100, each based on 100 observations, and he calculated the upper and lower bounds according to the leftmost and rightmost terms, respectively, of inequality (21.23). Figure 10.1 shows his results. He observed a decrease in the bounds with respect to n and obtained an upper bound of 1.3 bits per character for $n = 100$.

21.8   Relation of h, Perplexity, and Cross Entropy

The related parts in the main text appear in Sect. 10.5, page 110, and Sect. 16.1, page 164. The base of the logarithm in this section is 2.

A standard evaluation metric for language models is the perplexity (Manning and Schütze, 1999), which is a measure of the prediction accuracy. For clarity, let us consider a text sample consisting of a word sequence $w_1, \ldots, w_i, \ldots, w_m$, where $w_i \in W$ and the sequence's length is m. In general, the same argument applies for any other kind of element. Let $\hat{P}(w)$ denote the probability of a word w in a language model, which is acquired from a corpus.

The main text denotes the estimated probability given by a language model as P, but here, because we compare the *true* probability P with the "fake" probability of a language model that is *empirically acquired* by counting, we denote the former as P and the latter as $\hat{P}$, in this section only. Furthermore, the main text considers the complexity function $H(X)$ for a sequence X, but here, we want to use it to compare different probability functions, i.e., P and $\hat{P}$. Therefore, we denote it with the different notations $\mathbb{H}(P, \hat{P})$ and $\mathbb{H}(P)$ to indicate this difference.

The perplexity is defined as follows:

$$\text{perplexity} \equiv 2^{-\frac{1}{m} \sum_{i=1}^{m} \log \hat{P}(w_i)}. \tag{21.24}$$

The power term derives from the cross entropy between the true information source P (i.e., the original text) and the observation described by the language model $\hat{P}$, defined as follows:

$$\mathbb{H}(P, \hat{P}) \equiv - \sum_{w \in W} P(w) \log \hat{P}(w). \tag{21.25}$$

When $m \to \infty$, then the part $\dfrac{1}{m} \sum\limits_{i=1}^{m} \log \hat{P}(w_i)$ in the perplexity would go to
$\sum\limits_{w \in W} P(w) \log \hat{P}(w)$.

Next, let $\mathbb{H}(P) \equiv \mathbb{H}(P, P)$. The cross entropy and the entropy have the following relation (Cover and Thomas, 1991):

$$\mathbb{H}(P, \hat{P}) = \mathbb{H}(P) + \mathbb{KL}(P||\hat{P}). \qquad (21.26)$$

Here, $\mathbb{KL}$ is the Kullback–Leibler divergence, defined as follows:

$$\mathbb{KL}(P||\hat{P}) \equiv \sum_{w \in W} P(w) \log \frac{P(w)}{\hat{P}(w)}. \qquad (21.27)$$

Then, because $\mathbb{KL}(P||\hat{P}) \geq 0$ always, $\mathbb{H}(P, \hat{P}) \geq \mathbb{H}(P)$. Therefore,

$$\text{perplexity} \geq 2^h, \qquad (21.28)$$

where h is the true entropy rate. The perplexity, the metric to evaluate the model quality, is thus the upper bound of 2^h. In other words, estimation of h via some computation corresponds to acquiring the perplexity of a certain language model, as an upper bound.

21.9   Type Counts, Shannon Entropy, and Yule's K, via Generalized Entropy

The related part in the main text appears in Sect. 11.2, page 118. The base of the logarithm in this section is 2.

The relation between Harris's successor count and the Shannon entropy can be considered as follows. Let C be a set, with elements of it acquired as a variable Y. Let $P(Y = c)$ be the probability function that $c \in C$. For a constant $\rho \neq 1$, the Rényi complexity is defined as

$$H_\rho(Y) \equiv \frac{1}{1 - \rho} \log \sum_{c \in C} P^\rho(Y = c). \qquad (21.29)$$

This is one generalization of the Shannon entropy, and there are other generalizations such as the Tsallis entropy (Tsallis, 1988). A similar argument can be elaborated with the Tsallis entropy, but here we use the Rényi complexity.

The complexity is represented by H_ρ for different values of ρ, as follows. When $\rho = 0$,

$$H_0(Y) = \log |C|, \qquad (21.30)$$

because $P^0(Y = c) = 1$ for any element $c \in C$, so the summation counts the number of elements in the set C. When $\rho \to 1$, it is analytically proved that

$$H_\rho(Y) \Rightarrow H(Y), \tag{21.31}$$

i.e., the Shannon entropy (Rényi, 1961).

Furthermore, $H_2(Y)$ has been shown to correspond to the logarithm of Yule's K (Yule, 1944), when $P(Y = c)$ is considered via the relative frequency (Tanaka-Ishii and Aihara, 2015). Yule proposed K as an index that is invariant with respect to the text size. It represents the complexity of a text and has been used in the field of natural language processing. Yule's K is defined as follows. Let m be the total number of words in a text, and let v be the number of distinct words. Following Sect. 6.1, let $g(f)$ be the number of words appearing f times in the text. Finally, let $S_1 \equiv \sum_f fg(f)$ and $S_2 \equiv \sum_f f^2 g(f)$. Then,

$$K \equiv a \frac{S_2 - S_1}{S_1^2},$$

where a is a constant increasing the value of K, defined by Yule as $a = 10^4$. Reorganizing the term on the right easily shows its correspondence with $H_2(Y)$ in terms of the relative frequency.

Among $H_0(Y)$, $H(Y)$, and $H_2(Y)$, $H_0(Y)$ is known to increase infinitely with respect to the text length, as seen with the vocabulary growth in Chap. 6. $H_2(Y)$, with $P(Y = c)$ as the relative frequency, was empirically shown to converge to a value with increasing text length (Tanaka-Ishii and Aihara, 2015). In general, $P^\rho(Y)$ rapidly goes to zero for a larger ρ.

This difference in stability for the statistics of $H_\rho(Y)$ is one reason to reformulate Harris's hypothesis of articulation from an information-theoretic viewpoint.

21.10 Upper Bound of Compositional Distance

The related part in the main text appears in Sect. 12.5, page 131.

The proof by Tian et al. (2017) is pretty long, but a summary is provided here. For $1 \leq i \leq |W|$, let $f_i(t)$ be the frequency of the ith word to co-occur with a target word t, and let $p_i(t)$ be its probability. The ith element of $\vec{w}(t)$, the vector representation of word t, is defined as follows:

$$w_i \equiv F(f_i(t)) \equiv G(p_i(t)) - a(t) - b_i. \tag{21.32}$$

The function F appeared previously in the main text. The function G is one of several possible functions with which the following proof holds. Furthermore, $a(t)$ and b_i are defined so that the addition of the vector elements becomes zero and the center of all the word vectors becomes the origin.

As defined in the main text, let $s \backslash t$ denote the event in which t does not appear next to s, with st denoting otherwise. Similarly, let $\pi_{s \backslash t}$ denote the proportion of occurrences of s for which t does not appear next to s. The proof proceeds by considering the two cases in which t *does not* and *does* appear next to s. These cases correspond to the two terms on the right side of the following:

$$p_i(t) = \pi_{s \backslash t} p_i(s \backslash t) + (1 - \pi_{s \backslash t}) p_i(st). \tag{21.33}$$

A similar formula is deduced from this formula, as follows:

$$\vec{w}(t) \approx \pi_{s \backslash t} \vec{w}(s \backslash t) + (1 - \pi_{s \backslash t}) \vec{w}(st). \tag{21.34}$$

Corresponding formulas are obtained for $p_i(s)$ and $\vec{w}(s)$.

Next, the two terms on the left side of formula (12.4) are replaced with the formulas acquired above for $\vec{w}(t)$ and $\vec{w}(s)$. Then, by assuming $\vec{w}(st) \cdot \vec{w}(s \backslash t) \geq 0$, $\vec{w}(st) \cdot \vec{w}(t \backslash s) \geq 0$, and $\vec{w}(s \backslash t) \cdot \vec{w}(t \backslash s) \leq 0$, the right side of formula (12.4) is acquired.

21.11   Rough Summary of Mandelbrot's Communication Optimization Rationale to Deduce a Power Law

The related part in the main text appears in Sect. 15.1, page 156.

The rough mathematical rationale is as follows (Mandelbrot, 1965; Mitzenmacher, 2003). Partial differentiation of G by $P(r)$ gives the following:

$$\frac{\partial G}{\partial P(r)} = \frac{B(r)H + \text{Cost} \log_2(aP(r))}{H^2}, \tag{21.35}$$

where a is some constant. Setting this equal to zero and minimizing G yield

$$P(r) = \frac{2^{-\frac{HB(r)}{\text{Cost}}}}{a}. \tag{21.36}$$

The cost $B(r)$ of a word of rank r can be proved to be $B(r) \approx \log_u r$ as follows, where u is the number of characters. The rationale is similar to proving that monkey typing produces a power law in the rank-frequency distribution, as seen in Sects. 4.3 and 21.2. Let k be the length of the word for $B(r)$. Then, $B(r) = k + 1$ (i.e., add one for the space). The number of word types of length k is $(u - 1)^k$. As shorter words are more frequent, the rank for length k is $r = \sum_{i=1}^{k-1} (u - 1)^i$, thus giving $k \approx \log_{u-1} r$. Finally, substituting $B(r) \approx \log_{u-1} r$ in formula (21.36) gives

$$P(r) \propto r^{-\eta}, \tag{21.37}$$

where η is some constant.

Some readers might note that the partial differentiation should include the constraint that $P(r)$ adds up to 1, but Mitzenmacher (2003) confirmed $B(r)$ to be *within an additive constant of* $\log_{u-1} r$, so the distinction *can be ignored.*

21.12 Rough Definition of Central Limit Theorem

The related part in the main text appears in Sect. 15.2, page 158.

The central limit theorem is a statistical tool for data analysis. Consider m independent draws of $X_1, X_2, \ldots, X_m$ from the same information source, following a distribution with a finite mean μ and variance σ^2. This distribution does not have to be a normal distribution. Let $S_m = X_1 + \cdots + X_m$, and then, for $m \to \infty$,

$$\frac{S_m - m\mu}{\sigma\sqrt{m}} \Rightarrow N(0, 1), \tag{21.38}$$

where $N(0, 1)$ is the standard normal distribution. This indicates that the error between the sum S_m and the sum of the means, $m\mu$, which is normalized by $\sigma\sqrt{m}$, tends to follow a standard normal distribution when $m \to \infty$.

The description here is only a rough summary. The way of convergence is probabilistic, and the theorem has multiple variations. For the details, please refer to a standard textbook on statistics.

21.13 Definition of Simon Model

The related part in the main text appears in Sect. 16.3, page 166.

Let k_t be the number of kinds of elements (vocabulary size) at time t, and let $f_{t,i}$ be the frequency of elements of kind i occurring until t, where i is a positive integer (i.e., $i \in \mathbb{Z}_{\geq 1}$) at $t = 0$. A Simon model is initialized as follows:

$$k_0 = 1, \quad f_{0,1} = 1, \quad f_{0,j} = 0, \quad j \in \mathbb{Z}_{j>1}.$$

In other words, there is one element of type $i = 1$ in the sequence at $t = 0$ (i.e., there is only $f_{t=0,i=1}$, so $k_0 = 1$). For $t > 0$, given a constant $0 < a < 1$, an element is generated at time $t + 1$ with the following probabilities:

$$P(k_{t+1} = k_t + 1, f_{t+1,j} = f_{t,j}, j \in \mathbb{Z}_{\geq 1}\backslash\{k_t + 1\}, f_{t+1,k_t+1} = 1) \equiv a,$$

$$P(k_{t+1} = k_t, f_{t+1,i} = f_{t,i} + 1, f_{t+1,j} = f_{t,j}, j \in \mathbb{Z}_{\geq 1}\backslash\{i\})$$

$$\equiv (1 - a)\frac{f_{t,i}}{t}, \quad i = 1, \ldots, k_t.$$

Note that the first definition gives the case when a new word is introduced, while the second gives the case when a previous element is sampled according to the frequency distribution. For example, for the case of $X = [x, y, x, z, x, z]$, x, y, or z is sampled with probability $3(1 - a)/6$, $(1 - a)/6$, or $2(1 - a)/6$, respectively. This reuse of previous elements according to the frequency distribution is equivalent to sampling uniformly from the past sequence. Thus, in this example, uniform sampling entails picking one element randomly from $X = [x, y, x, z, x, z]$.

Chapter 22
Data

The analyses in this book use a large quantity of data, all of which is publicly available from data resource consortia. The data is categorized into four types: single-author literary texts; large, multi-author corpora; miscellaneous data related to language; and corpora for analyzing different scripts. This chapter describes these datasets and the preprocessing applied to them.

22.1 Literary Texts

Table 22.1 lists the literary texts that were mainly used in Parts II and III. The natural language texts consisted of 1142 long, single-author texts in 14 languages from Project Gutenberg and Aozora Bunko. The last row lists the statistics for *Moby Dick*.

The size threshold was 1 MB, including annotations. Project Gutenberg splits many of the literary texts into different volumes, so they were manually combined into single texts.

Once the texts were collected, the metadata and annotations were removed so that only the original texts remained. The Pynlpir and Mecab applications were used to segment the Chinese and Japanese texts, respectively. NLTK was used to tokenize the texts of other languages.

Defining the word set was often difficult. Lemmatization also requires an accurate lemmatizer, which is not available for some non-major languages. In addition, processing of upper and lower case letters, misspellings, and symbols presents problems. It is difficult to consistently handle these aspects and problems across different writing systems. As the main results in this book do not change much whether words are lemmatized or not,[1] it was decided that a fair approach

---

[1] Zipf (1949) also mentioned this.

© The Author(s) 2021
K. Tanaka-Ishii, *Statistical Universals of Language*, Mathematics in Mind,
https://doi.org/10.1007/978-3-030-59377-3_22

Table 22.1 Summary of the literary texts used in this book. For each dataset, the length is the total number of words, and the vocabulary is the number of different words

Text source	Language	Number of samples	Length			Vocabulary		
			Mean	Min	Max	Mean	Min	Max
Gutenberg	English	910	313127.5	185939	2488933	17236.7	7320	69811
	French	66	293192.5	169415	1528177	22097.4	14105	57192
	Finnish	33	197519.3	149488	396920	33596.2	26274	47262
	Chinese	32	629916.8	315099	4145117	15352.9	9152	60949
	Dutch	27	256859.3	198924	435683	19158.1	13879	31594
	German	20	236174.0	184320	331321	24241.3	11078	37227
	Italian	14	266809.7	196961	369326	29102.6	18640	45031
	Spanish	12	363837.2	219787	903051	26110.1	18110	36506
	Greek	10	159969.2	119196	243953	22804.7	15876	31385
	Latin	2	505743.5	205228	806259	59666.5	28738	90595
	Portuguese	1	261382.0	261382	261382	24718.0	24718	24718
	Hungarian	1	198303.0	198303	198303	38383.0	38383	38383
	Tagalog	1	208455.0	208455	208455	26334.0	26334	26334
Aozora	Japanese	13	616676.2	105342	2951319	19759.0	6619	49099
Moby Dick	English	1	254654			20472		

Table 22.2 Summary of the large data used in this book. For each dataset, the length is the total number of words, and the vocabulary is the number of different words

Corpus	Language	Length	Vocabulary
Web-E	English	88267343947	204724587
Web-J	Japanese	7183558565	5474644
Penn Treebank	English	892008	89317
Wall Street Journal	English	22679512	137466
New York Times (in Gigaword Corpus)	English	1528137194	3155494

would be to use all tokenized results, without introducing any arbitrary elimination scheme. In other words, at the word level, conjugated words, capitalized words, misspelled words, and symbols were included in the sequences used to produce the results in this book. This preprocessing scheme was also applied to the corpus data described below.

22.2   Large Corpora

Table 22.2 lists the large-scale corpora that were studied in this book. In contrast to the literary texts, these corpora are characterized as multi-author texts.

Chapter 12 uses Internet data collected in 2006, which is listed in the first two rows of Table 22.2. The analysis of familiarity required data with gigantic coverage. On the other hand, these web page collections were inadequate for verifying statistical universals in other chapters that required a text consisting of sentences, as such web data often contains other linguistic usages besides sentences.

Often, very large-scale corpora, even those distributed by reputable organizations, have problems such as long repetitions, chunks in different languages, sequences of numbers, and large numbers of spaces. Of particular issue are repetitions, because statistical analysis involves analyzing counts. Arbitrary long repetitions seem to include two cases: intentional and unintentional (arising from some procedural error). The former case includes copyright claims, for example.

Another issue is the language. Even if a corpus is in one language, it could have been produced in a nonnative country (for example, the corpus of Newspapers of Japan in English, created in Japan). In such corpora, many keywords appear in a different language than that of the main text, and there are some questions as to whether these corpora are useful for studying the nature of a particular language.

As mentioned at the beginning of Chap. 3, a corpus is not just an accumulation of texts; it is composed by applying rules to the source data. The data sources in this book were chosen after considering their adequacy for highlighting certain procedures. In particular, most of the analyses on large data were conducted on standard, clean newspaper corpora.

The *Wall Street Journal* has been used in many studies in computational linguistics. In particular, a grammatically annotated portion of the *Wall Street Journal*, the Penn Treebank, is the gold standard corpus of this type. It has also been used to analyze neural models. The third row of Table 22.2 lists statistics for the Penn Treebank.

As shown in the table, however, the Penn Treebank is rather smaller than the *Wall Street Journal*: it is too small for some analyses. Moreover, as the *Wall Street Journal* is mainly composed of financial articles, its vocabulary may be too focused for some analyses. For this reason, the analyses in Parts III and IV use instead a corpus made from articles of the *New York Times*. Its data is deemed the cleanest among other English corpora included in the English Gigaword Corpus distributed by the Language Data Consortium. The last row of Table 22.2 lists this corpus.

22.3   Other Kinds of Data Related to Language

Some chapters of this book examine the behavior of statistical universals in various kinds of data other than text. Table 22.3 summarizes this data. The corpora listed in the lower block are analyzed in Chap. 5. Those in the upper block are used in Chap. 9 to reveal the general effects of long memory.

The first row of Table 22.3 lists the enwiki8 100-MB dump dataset, consisting of tag-annotated text from English Wikipedia. This dataset, which includes annotations by Wikipedia, was used to study the effect of annotations.

Obtaining enough clean speech data for the analyses presented in Parts II and III is difficult. Recording sessions tend to be very short because the content of speech tends to change quickly. Various sources of speech data were examined

Table 22.3 Summary of the language-related data used in this book. For each dataset, the length is the total number of words, and the vocabulary is the number of different words

Text source	Language	Number of samples	Length			Vocabulary		
			Mean	Min	Max	Mean	Min	Max
Tagged Wiki	Tag-annotated (enwiki8)	1	14647847			1430790		
Diet record	Japanese	250	348902.8	101441	1467337	9511.4	4957	20429
CHILDES	Various	10	193434.0	48952	448772	9907.0	5618	17892
Program source code	Various	4	3461017.8	3697198	68622161	838906.8	127652	1545126
Music	MIDI	12	135992.4	76628	215479	9186.9	906	27042
TED talk	English	1	10770			1680		
Thomas	English	1	448772			17892		
Haskell	Haskell	1	68622161			1545126		
Beethoven's 9th	MIDI	1	215479			27042		

as candidates for the analyses presented in this book, and two sources were chosen. The first is the spoken record of 250 sessions of Japan's National Diet (the second row of Table 22.3 lists its statistics). Each session was an opening of a National Diet meeting. The speech data transcribed by professionals is clean and long. The second source is a transcript of a TED talk in English.[2] Although this is one of the longest TED talks, the statistics show that it is still rather short for our purposes. For that reason, the Diet records were used for the analysis of long memory in Part III.

This book also considers CHILDES (Child Language Data Exchange System) data, as listed in the third row of the table.[3] The 10 longest child-directed speech utterances in the CHILDES database were used (Thomas (MacWhinney, 2000; Lieven et al., 2009), Groningen (Bol, 1995), Rondal (Rondal, 1985), Leo (Behrens, 2006), Ris (Gil and Tadmor, 2007), Nanami (Oshima-Takane and MacWhinney, 1995), Inka (Smoczyńska, 1985), Angela (Andelkovic et al., 2001), Beca (Benedet et al., 2004), and Boteborg (Plunkett and Strömqvist, 1992)). The data was preprocessed by extracting only the children's utterances from the recordings. In the lower block, the table lists the data for one child, Thomas, who was recorded in different sessions from when he was two to four years old. Thomas is a native English speaker, and his record is the longest in the CHILDES data.

The fourth row of the table lists program source code data (in the Lisp, Haskell, C++, and Python programming languages) that was crawled from large representative archives. The data was parsed and stripped of comments written in natural language. The data for Haskell, used in Chap. 5, appears in the third row of the table's lower block. This data includes many repetitions because of the tendency to copy and paste sample code. The results presented in Chap. 9 should only be viewed as one possible result; it remains an open question how best to analyze program source code as text data.

The statistics for music in the table were taken from 12 pieces (long symphonies and so forth). They were transformed from MIDI data into text with the software SMF2MML,[4] and then the annotations were removed. SMF2MML transcribes every sound and its duration by using a descriptive method. The transcriptions thus could be treated as sets of words. The analyses presented in Chap. 5 and Chap. 9 require a large amount of data. For this reason, only long classical music pieces, such as symphonies, were chosen. The last row of the lower block lists the statistics of a sample of Beethoven's Symphony No. 9, which is analyzed in Chap. 5. As is the case with the program source code data, there are various other preprocessing techniques that can be applied to music data. The results shown in Chap. 5 and Chap. 9 should thus be regarded as those of only one possible analysis.

---

[2]The TED talk was *Nationalism vs. Globalism, The New Political Divide* by Yuval Noah Harari, transcribed in English.

[3]I hereby thank Prof. Franklin Chang of Kobe City University of Foreign Studies, a psycholinguist knowledgeable on CHILDES data analysis, for his guidance on this dataset.

[4]http://shaw.la.coocan.jp/smf2mml/.

Table 22.4 Summary of the datasets used for different scripts in this book. For each dataset, the table lists the length and size in characters

Text source	Language	Length	Script size
Wall Street Journal	English	112868099	52
Watan corpus	Arabic	42174262	36
Hindi corpus	Hindi	12921901	80
People's Daily	Chinese	32189125	5960
Mainichi	Japanese	56395393	4755
KAIST corpus	Korean	28372148	6237

22.4 Corpora for Scripts

Table 22.4 lists the corpora considered for the writing scripts in Chap. 5. Daniels and Bright (1996) categorized scripts around the globe into six categories—alphabet, abjad, abugida, logosyllabary, syllabary, and featural—and Chap. 5 uses English, Arabic, Hindi, Chinese, Japanese, and Korean, respectively, as representatives. A corpus was chosen for each representative language. Note that the Japanese kana script is a syllabary, but the Japanese script also includes logosyllabic characters adapted from Chinese. To date, there is no naturally produced large-scale corpus of quality for only the syllabary, so the full Japanese script is used instead.

The same preprocessing was applied to all the corpora. All metadata was removed and text was extracted before the characters were separated. The definitions of the characters in each script are based on Unicode.[5] Therefore, the analysis in Chap. 5 *excludes* all symbols, such as spaces, punctuation, and so on, and was conducted only on the set of characters for each language.

---

[5]This was implemented with the Python library `unicode_script`.

Correction to: Statistical Universals of Language

Kumiko Tanaka-Ishii

Correction to:
K. Tanaka-Ishii, *Statistical Universals of Language*,
Mathematics in Mind,
https://doi.org/10.1007/978-3-030-59377-3

Chapter 3
The original version of the chapter "Language as a Complex System" was previously published without updating missing reference in footnote 4, page 22. This change has now been included and the chapter and the book have been updated with the change.

Chapter 4
The original version of the chapter "Relation Between Rank and Frequency" was previously published without including the updated modified text on page 43, 1st para. This change has now been included and the chapter and the book have been updated with the change.

Chapter 6
The original version of the chapter "Related Statistical Universals" was previously published without including the updated equations on pages 57 (3rd para, 8th line), 60 (last para, 3rd line) and 61(1st para, 1st line). These changes have now been included and the chapter and the book have been updated with the change.

Chapter 8
The original version of the chapter "Long-Range Correlation" was previously published without including the updated modified text on page 84, 3rd para under Section 8.5. This change has now been included and the chapter and the book have been updated with the change.

Chapter 14
The original version of the chapter "Grammatical Structure and Long Memory" was previously published without including the end period on page 144, line 27, and page 145, line 2. These changes have now been included and the chapter and the book have been updated with the change.

Chapter 15
The original version of the chapter "Theories Behind Zipf's Law" was previously published without including the updated modified text on page 156, line 8. This change has now been included and the chapter and the book have been updated with the change.

References
The original version of the "References" was previously published without including the updated reference details for "Li, Wentian (1989)" on page 227. This change has now been included and the References and the book have been updated with the change.

The updated original version of this chapter can be found at
https://doi.org/10.1007/978-3-030-59377-3_3
https://doi.org/10.1007/978-3-030-59377-3_4
https://doi.org/10.1007/978-3-030-59377-3_6
https://doi.org/10.1007/978-3-030-59377-3_8
https://doi.org/10.1007/978-3-030-59377-3_14
https://doi.org/10.1007/978-3-030-59377-3_15
https://doi.org/10.1007/978-3-030-59377-3

Correction to: Glossary and Notations

Correction to:
K. Tanaka-Ishii, *Statistical Universals of Language*,
Mathematics in Mind,
https://doi.org/10.1007/978-3-030-59377-3_20

The original version of this chapter was published with the wrong text "logarithmic base is 10" on page 202, line 17. It has been updated with the correct text as "natural logarithmic base is used" in the chapter.

The updated original version of this chapter can be found at
https://doi.org/10.1007/978-3-030-59377-3_20

Correction to: Statistical Universals of Language

Kumiko Tanaka-Ishii

Correction to:
K. Tanaka-Ishii, *Statistical Universals of Language*,
Mathematics in Mind,
https://doi.org/10.1007/978-3-030-59377-3

Chapter 1
The original version of this chapter was published with the wrong text "de Saussure (1916)," on page 8, line 3. Now, the correct text "de Saussure (1911)," has been updated in the chapter.

References
The original version of this chapter was published without updating the reference "de Saussure, Ferdinand" on page 225. Now, the correct reference has been updated in the chapter.

The updated original version of these chapters can be found at
https://doi.org/10.1007/978-3-030-59377-3_1
https://doi.org/10.1007/978-3-030-59377-3

References

Aho, Alfred V., Lam, Monica S., Sethi, Ravi, and Ullman, Jeffrey D. (1986). *Compilers : Principles, Techniques, and Tools*. Addison-Wesley.

Aitchison, Jean (1987). *Words in the mind: an introduction to the mental lexicon*. Basil Blackwell Ltd.

Allahverdyan, Armen E., Deng, Weibing, and Wang, Qiuping A. (2013). Explaining Zipf's law via a mental lexicon. *Physical Review E*, **88**:062804.

Altmann, Eduardo G. and Gerlach, Martin (2016). Statistical laws in linguistics. *Creativity and Universality in Language*, pages 7–26.

Altmann, Eduardo G., Pierrehumbert, Janet B., and Motter, Adilson E. (2009). Beyond word frequency: Bursts, lulls, and scaling in the temporal distributions of words. *PLoS One*, **4**(11):e7678.

Altmann, Edouard G., Cristadoro, Giampaolo and Esposti, Mirko D. (2012). On the origin of long-range correlations in texts. *Proceedings of the National Academy of Sciences*, **109**(29), 11582–11587.

Amano, Shigeaki and Kondo, Kimihisa (2000). On the NTT psycholinguistic databases : Lexical properties of Japanese. *Journal of the Phonetic Society of Japan*, **4**(2), 44–50.

Andelkovic, Darinka, Seva, Nada, and Moskovljevic, Jasmina (2001). *Serbian Corpus of Early Child Language*. Laboratory for Experimental Psychology, Faculty of Philosophy, and Department of General Linguistics, Faculty of Philology, University of Belgrade.

Baayen, R. Harald (2001). *Word Frequency Distributions*. Springer.

Baayen, R. Harald and Lieber, Rochelle (1996). Word frequency distributions and lexical semantics. *Computers and the Humanities*, **30**, 281–291.

Badii, Remo and Politi, Antonio (1997). *Complexity: Hierarchical structures and scaling in physics*. Cambridge University Press.

Baeza-Yates, Ricard and Navarro, Gonzalo (2000). Block addressing indices for approximate text retrieval. *Journal of the American Society for Information Science*, **51**, 69–82.

Baixeries, Jaume, Brita, Elvevag, and Ferrer-i-Cancho, Ramon (2013). The evolution of the exponent of Zipf's law in language ontogeny. *PLoS One*, **8**(3):e53227

Bak, Per, Tang, Chao, and Wiesenfeld, Kurt (1987). Self-organized criticality: An explanation of the 1/f noise. *Physical Review Letters*, **59**(4), 381–384.

Bak, Per, Tang, Chao, and Wiesenfeld, Kurt (1988). Self-organized criticality. *Physical Review A*, **38**(1), 364–374.

© The Author(s) 2021
K. Tanaka-Ishii, *Statistical Universals of Language*, Mathematics in Mind,
https://doi.org/10.1007/978-3-030-59377-3

Barabasí, Albert-Laszlo and Albert, Reka (1999). Emergence of scaling in random networks. *Science*, **286**, 509–512.

Barabási, Albert-László (2016). *Network Science*. Cambridge University Press.

Beaney, Michael (1997). *The Frege Reader*, pages 151–180. Blackwell Publishing.

Behrens, Heike (2006). The input-output relationship in first language acquisition. *Language and Cognitive Processes*, **21**, 2–24.

Bell, Timothy C., Cleary, John G., and Witten, Ian H. (1990). *Text Compression*. Prentice Hall.

Benedet, Maria, Cruz, Celis, Carrasco, Maria, and Snow, Catherine (2004). *Spanish BecaCESNo Corpus*. TalkBank.

Bengio, Yoshua, Simard, Patrice Y., and Frasconi, Paolo (1994). Learning long-term dependencies with gradient descent is difficult. *IEEE Transactions on Neural Networks*, **5**, 157–166.

Bentz, Christian and Ferrer-i-Cancho, Ramon (2016). Zipf's law of abbreviation as a language universal. In *Proceedings of the Leiden Workshop on Capturing Phylogenetic Algorithms for Linguistics*.

Berger, Toby (1968). Rate distortion theory for sources with abstract alphabets and memory. *Information and Control*, **13**, 254–273.

Bernhardsson, Sebastian, da Rocha, Luis E. C., and Minnhagen, Petter (2009). The meta book and size-dependent properties of written language. *New Journal of Physics*, **11**(12):123015.

Blender, Richard, Raible, Christoph C., and Lunkeit, Frank (2014). Non-exponential return time distributions for vorticity extremes explained by fractional poisson processes. *Quarterly Journal of the Royal Meteorological Society*, **141**, 249–257.

Blevins, James P. and Sag, Ivan A. (2013). *Phrase structure grammar*, chapter 7, pages 202–225. Cambridge University Press. Editor den Dikken, Marcel.

Bogachev, Mikhail I., Eichner, Jan F., and Bunde, Armin (2007). Effect of nonlinear correlations on the statistics of return intervals in multifractal data sets. *Physical Review Letters*, **99**(24):240601.

Bojanowski, Piotr, Grave, Edouard, Joulin, Armand, and Mikolov, Tomas (2017). Enriching word vectors with subword information. *Transactions of the Association for Computational Linguistics*, **5**, 135–146.

Bol, Gerard W. (1995). Implicational scaling in child language acquisition : the order of production of Dutch verb constructions. In *The Dutch-German Colloquium on Language Acquisition*, Amsterdam Series in Child Language Development, pages 1–13. Amsterdam: Institute for General Linguistics. edited by Verrips, M. and Wijnen, F.

Brown, Peter F., Della-Pietra, Stephan A., Della-Pietra, Vincent J., Lai, Jennifer C., and Mercer, Robert L. (1992). An estimate of an upper bound for the entropy of English. *Computational Linguistics*, **18**(1), 31–40.

Buchholz, Sabine N. (2002). *Memory-Based Grammatical Relation Finding*. Eigen beheer. Doctoral Thesis.

Bunde, Armin and Havlin, Shlomo (1996). *Fractals and Disordered Systems*. Springer.

Bunde, Armin, Eichner, Jan F., Kantelhardt, Jan W., and Havlin, Shlomo (2005). Long-term memory : A natural mechanism for the clustering of extreme events and anomalous residual times in climate records. *Physical Review Letters*, **94**(4):048701.

Chomsky, Noam (1957). *Syntactic Structures*. Mouton & Co.

Chomsky, Noam (1965). *Aspects of the theory of syntax*. The MIT Press.

Chomsky, Noam (1995). *The Minimalist Program*. The MIT Press.

Christiansen, Morten H., Collins, Chrsitopher, and Edelman, Shimon (2009). *Language Universals*. Oxford University Press.

Clauset, Aaron, Shalizi, Cosma R., and Newman, Mark E. J. (2009). Power-law distributions in empirical data. *SIAM review*, **51**(4), 661–703.

Comrie, Bernard (1981). *Language Universals and Linguistic Typology: Syntax and Morphology*. The University of Chicago Press.

Connine, Cynthia M., Mullennix, John, Shernoff, Eve, and Yelen, Jennifer (1990). Word familiarity and frequency in visual and auditory word recognition. *Journal of Experimental Psychology: Learning Memory and Cognition*, **16**(6), 1084–1096.

Conrad, Brian and Mitzenmacher, Michael (2004). Power laws for monkeys typing randomly: The case of unequal probabilities. *IEEE Transactions on Information Theory*, **50**(7), 1403–1414.

Corral, Álvaro (2004). Long-term clustering, scaling, and universality in the temporal occurrence of earthquakes. *Physical Review Letters*, **92**(10):108501.

Corral, Álvaro (2005). Renomalization-group transformations and correlations of seismicity. *Physical Review Letters*, **95**:028501.

Coulmas, Florian (1996). *The Blackwell Encyclopedia of Writing Systems*. Blackwell Publishers Ltd.

Cover, Thomas M. and King, Roger C. (1978). A convergent gambling estimate of the entropy of English. *IEEE Transactions on Information Theory*, **24**(4), 413–421.

Cover, Thomas M. and Thomas, Joy A. (1991). *Elements of Information Theory*. John Wiley & Sons, Inc.

Creutz, Mathias and Lagus, Krista (2002). Unsupervised discovery of morphemes. In *Proceedings of the ACL-02 Workshop on Morphological and Phonological Learning*, pages 21–30.

Crutchfield, J. P. and Feldman, D. P. (2003). Regularities unseen, randomness observed: Levels of entropy convergence. *Chaos*, **13**, 25–54.

Dai, Zihang, Yang, Zhilin, Yang, Yiming, Carbonell, Jaime, Le, Quoc V., and Salakhutdinov, Ruslan (2019). Transformer-XL: Attentive language models beyond a fixed-length context. In *Proceedings of the 57th Annual Meeting of the Association for Computational Linguistics*, pages 2978–2988.

Daniels, Peter T. and Bright, William, editors (1996). *The World's Writing Systems*. Oxford University Press.

Dębowski, Łukasz (2015). The relaxed Hilberg conjecture: A review and new experimental support. *Journal of Quantitative Linguistics*, **22**(4), 311–337.

Dębowski, Łukasz (2020). *Information Theory Meets Power Laws: Stochastic Processes and Language Models*. Wiley.

de Saussure, Ferdinand (1911). Troisième Cours de Linguistique Générale (1910 - 1911). From the notebooks of Emile Constantin. Pergamon Press. English translation by Harris, Roy, 1993.

Deng, Weibing, Allahverdyan, Armen E., Li, Bo, and Wang, Quipiing A. (2014). Rank-frequency relation for Chinese characters. *The European Physical Journal B*, **87**:47.

Dryer, Matthew S. (1988). Object-verb order and adjective-noun order: dispelling a myth. *Lingua*, **74**, 185–217.

Dryer, Matthew S. (1992). The Greenbergian word order correlations. *Language*, **68**(1), 81–138.

Dupoux, Emmanuel and Mehler, Jacques (1990). Monitoring the lexicon with normal and compressed speech : Frequency effects and the prelexical code. *Journal of Memory and Language*, **29**, 316–335.

Ebeling, Werner and Neiman, Alexander (1995). Long-range correlations between letters and sentences in texts. *Physica A*, **215**, 233–241.

Ebeling, Werner and Nicolis, G. (1991). Entropy of symbolic sequences : The role of correlations. *Europhysics Letters*, **14**(3), 191–196.

Ebeling, Werner and Pöschel, Thorsten (1993). Entropy and long-range correlations in literary English. *Europhysics Letters*, **26**(4), 241–246.

Eisler, Zoltán, Bartos, Imre, and Kertész, János (2008). Fluctuation scaling in complex systems: Taylor's law and beyond. *Advances in Physics*, **57**, 89–142.

Fechner, Gustav T. (1860). *Elements of psychophysics*, volume 1. Holt, Rinehard and Winston. *Elemente der Psychophysik* Eds. Howes, D.H., Boring, E.G., translated by H.E.Adler, published in 1966.

Fernández-González, Daniel and Martins, André F. T. (2015). Parsing as reduction. In *Proceedings of the 53rd Annual Meeting of the Association for Computational Linguistics and the 7th International Joint Conference on Natural Language Processing Volume 1 : Long Papers*, pages 1523–1533. Association for Computational Linguistics.

Ferrer-i-Cancho, Ramon (2018). Optimization models of natural communication. *Journal of Quantitative Linguistics*, **25**(3), 207–237.

Ferrer-i-Cancho, Ramon and Elvevåg, Brita (2010). Random texts do not exhibit the real Zipf's law-like rank distribution. *PLoS ONE*, **5**(3). e9411.

Ferrer-i-Cancho, Ramon, Dębowski, Łukas, and Moscoso del Prado Martin, Fermin **2013**:L07001. Constant conditional entropy and related hypotheses. *Journal of Statistical Mechanics: Theory and Experiment*, **2013**.

Font-Clos, Francesc and Corral, Álvaro (2015). Log-log convexity of type-token growth in Zipf's systems. *Physical Review Letters*, **114**:238701.

Foucault, Michel (1969). *L'Archéologie du Savoir*. Editions Gallimard.

Frantzi, Katerina T. and Ananiadou, Sophia (1996). Extracting nested collocations. In *Proceedings of the 16th International conference on Computational linguistics*, pages 41–46.

Frege, Gottlob (1892). *Über Sinn und Bedeutung*, pages 25–50. Zeitschrift für Philosophie und Philosophische Kritik, Vol. 100.

Genzel, Dmitriy and Charniak, Eugene (2002). Entropy rate constancy in text. In *Proccedings of the 40th Annual Meeting on Association for Computational Linguistics*, pages 199–206.

Gerlach, Martin and Altmann, Eduardo G. (2013). Stochastic model for the vocabulary growth in natural languages. *Physical Review X*, **3**(2):21006.

Gerlach, Martin and Altmann, Eduardo G. (2014). Scaling laws and fluctuations in the statistics of word frequencies. *New Journal of Physics*, **16**:113010.

Gil, David and Tadmor, Uri (2007). *The MPI-EVA Jakarta Child Language Database*. A joint project of the Department of Linguistics, Max Planck Institute for Evolutionary Anthropology and the Center for Language and Culture Studies, Atma Jaya Catholic University.

Goldwater, Sharon, Griffiths, Thomas L., and Johnson, Mark (2009). A Bayesian framework for word segmentation: Exploring the effects of context. *Cognition*, **112**, 21–54.

Good, Ian J. (1953). The population frequencies of species and the estimation of population parameters. *Biometrika*, **40**, 237–264.

Grave, Edouard, Joulin, Armand, and Usunier, Nicolas (2017). Improving neural language models with a continuous cache. In *Proceedings of International Conference on Learning Representations*.

Greenberg, Joseph H. (1963). *Universals of Language*. The MIT Press. In the chapter entitled "Some universals of grammar with particular reference to the order of meaningful elements", pages 73–113.

Grice, Paul (1989). *Studies in the Way of Words*. Harvard University Press.

Guiraud, Pierre (1954). *Les Caractères Statistique du Vocabulaire*. Universitaires de France Press.

Hafer, Margaret A.and Weiss, Stefan F. (1974). Word segmentation by letter successor varieties. *Information Storage and Retrieval*, **10**, 371–385.

Harris, Zellig S. (1954). Distributional structure. *Word*, **10**(2–3), 146–162.

Harris, Zellig S. (1955). From phoneme to morpheme. *Language*, **31**(2), 190–222.

Harris, Zellig S. (1968). *Mathematical Structures of Language*. Interscience Publishers (John Wiley & Sons).

Harris, Zellig S. (1988). *Language and Information*. Columbia University Press.

Haspelmath, Martin, Dryer, Matthew S., Gil, David, and Comrie, Bernard (2005). *The World Atlas of Language Structures*. Oxford University Press.

Hawkins, John A. (1990). A parsing theory of word order universals. *Linguistic Inquiry*, **21**(2), 223–261.

Heaps, Harold S. (1978). *Information Retrieval: Computational and Theoretical Aspects*. Academic Press.

Heidelberger, Michael (1993). *Nature from Within: Gustav Theodor Fechner and His Psychophysical Worldview*. Vittorio Klostermann GmbH.

Herdan, Gustav (1956). *Language as Choice and Chance*. Noordhoff.

Herdan, Gustav (1964). *Quantitative Linguistics*. Butterworths.

Hey, Tony, Tansley, Stewart, and Tolle, Kristin (2009). *The Fourth Paradigm: Data-Intensive Scientific Discovery*. Microsoft Research.

Hilberg, Wolfgang (1990). Der bekannte grenzwert der redundanzfreien information in texten — eine fehlinterpretation der shannonschen experimente? *Frequenz*, **44**, 243–248.

Hochreiter, Sepp and Schmidhuber, Jürgen (1997). Long short-term memory. *Neural Computation*, **9**(8), 1735–1780.

Hockett, Charles F. (1958). *A course in modern linguistics*. Macmillan Company.

Huang, Jin-Hu and Powers, David (2003). Chinese word segmentation based on contexual entropy. *Proceedings of the 17th Pacific Asia Conference on Language, Information and Computation*, pages 152–158.

Hurst, Harold E. (1951). Long-term storage capacity of reservoirs. *Transactions of the American Society of Civil Engineers*, **116**(11), 770–799.

Jensen, Henrik J. (1998). *Self-Organized Criticality: Emergent Complex Behavior in Physical and Biological Systems*. Cambridge University Press.

Kamp, Hans and Reyle, Uwe (1993). *From Discourse to Logic*. Springer.

Kantelhardt, Jan W., Zschiegner, Stephan A.and Koscielny-Bunde, Eva, Havlin, Shlomo, Bunde, Armin, and Stanley, H. Eugene (2002). Multifractal detrended fluctuation analysis of nonstationary time series. *Physica A*, **316**, 87–114.

Kanwal, Jasmeen, Smith, Kenny, Culbertson, Jennifer, and Kirby, Simon (2017). Zipf's law of abbreviation and the principle of least effort: Language users optimise a miniature lexicon for efficient communication. *Cognition*, **165**, 45–52.

Katz, Slava M. (1987). Estimation of probabilities from sparse data for the language model component of a speech recognizer. *IEEE Transactions on Acoustics, Speech, and Signal Processing*, **35**(3), 400–401.

Kempe, André (1999). Experiments in unsupervised entropy-based corpus segmentation. In *EACL 1999: CoNLL-99 in Computational Natural Language Learning*, pages 7–13.

Kneser, Reinhard and Ney, Hermann (1995). Improved backing-off for n-gram language modeling. In *1995 International Conference on Acoustics, Speech, and Signal Processing*, volume 1, pages 181–184.

Kobayashi, Tatsuru and Tanaka-Ishii, Kumiko (2018). Taylor's law for human linguistic sequences. *Proceedings of the 56th Annual Meeting of the Association for Computational Lingusitics*, pages 1138–1148.

Kong, Lingpeng, Rush, Alexander M., and Smith, Noah A. (2015). Transforming dependencies into phrase structures. *Proceedings of the 2015 Annual Conference of the North American Chapter of the Association for Computational Linguistics*, pages 788–798.

Kosmidis, Kosmas, Kalampokis, Alkiviadis, and Argyrakis, Panos (2006). Language time series analysis. *Physica A*, **370**, 808–816.

Krizhevsky, Alex, Sutskever, Ilya, and Hinton, Geoffrey E. (2012). ImageNet classification with deep convolutional neural networks. In *the Proceedings of the 25th International Conference on Neural Information Processing Systems*, volume 1, pages 1097–1105.

Kretzschmar Jr., William A. (2015) *Language and Complex Systems*. Cambridge University Press.

Lanczos, Cornelius (1949). *The Variational Principles of Mechanics*. University of Toronto Press.

Lennartz, Sabine and Bunde, Armin (2009). Eliminating finite-size effects and detecting the amount of white noise in short records with long-term memory. *Physical Review E*, **79**(6):066101.

Levy, Roger and Jaeger, T. Florian (2006). Speakers optimize information density through syntactic reduction. In *Advances in Neural Information Processing Systems 19*, pages 849–856. MIT Press.

Li, Wentian (1989). Mutual information functions of natural language texts. *Santa Fe Institute Working Paper*, 1989.

Li, Wentian (1992). Random texts exhibit Zipf's-law-like word frequency distribution. *IEEE Transactions on Information Theory*, **38**, 1842–1845.

Li, Wentian, Marr, Thomas G., and Kaneko, Kunihiko (1994). Understanding long-range correlations in DNA sequences. *Physica D : Nonlinear Phenomena*, **75**, 392–416.

Lieven, Elena, Salomo, Dorothé, and Tomasello, Michael (2009). Two-year-old children's production of multiword utterances : A usage-based analysis. *Cognitive Linguistics*, **20**(3), 481–507.

Lin, Henry W. and Tegmark, Max (2017). Critial behavior in physics and probabilistic formal languages. *Entropy*, **19**(7):299.

Lu, Ang, Wang, Weiran, Bansal, Mohit, Gimpel, Kevin, and Livescu, Karen (2015). Deep multilingual correlation for improved word embeddings. *Proceedings of the 2015 Conference of the North American Chapter of the Association for Computational Linguistics: Human Language Technologies*, pages 250–256.

Lü, Linyuan, Zhang, Zi-Ke, and Zhou, Tao (2010). Zipf's law leads to Heaps' law : Analyzing their relation in finite-size systems. *PLoS ONE*, **5**(12):e14139.

Lü, Linyuan, Zhang, Zi-Ke, and Zhou, Tao (2013). Deviation of Zipf's and Heaps' laws in human languages with limited dictionary sizes. *Scientific Reports*, **1082**.

MacWhinney, Brian (2000). *The Childes Project*. Lawrence Erlbaum Associates, Inc.

Mallarmé, Stéphane (1897). *Un Coup de Dés and Other Poems*. Poetry In Translation. Un coup de dés jamais n'abolira le hasard, Translation by A. S. Kline.

Mandelbrot, Benoit B. (1953). An informational theory of the statistical structure of language. In *Proceedings of Symposium of Applications of Communication theory*, pages 486–502.

Mandelbrot, Benoit B. (1965). *Information Theory and Psycholinguistics. Scientific Psychology*, pages 250—368.

Mandelbrot, Benoit B. (1997). *Fractals and Scaling in Finance: Discontinuity, Concentration, Risk.* Springer-Verlag.

Manning, Christopher D. and Schütze, Hinrich (1999). *Foundations of Statistical Natural Language Processing*. The MIT Press.

Marcus, Mitchell, Kim, Grace, Marcinkiewicz, Mary A., Macintyre, Robert, Bies, Ann, Ferguson, Mark, Katz, Karen, and Schasberger, Britta (1994). The Penn Treebank: Annotating predicate argument structure. *HLT'94 Proceedings of the Workshop on Human Language Technology*, volume 6, pages 114–119.

Marcus, Mitchell P., Santorini, Beatrice, and Marcinkiewicz, Mary Ann (1993). Building a large annotated corpus of English: the Penn Treebank. *Computational Linguistics*, **19**(2), 313–330.

Marslen-Wilson, William D. (1990). Activation, Competition, and Frequency in Lexical Access. *Cognitive Models of Speech Processing*, pages 148–172. The MIT Press.

Martinet, André (1960). *Eléments de linguistique générale*. Armand Colin.

Merity, Stephen, Keskar, Nitish S., and Socher, Richard (2018). An analysis of neural language modeling at multiple scales. *CoRR*, **abs/1803.08240**.

Mikolov, Tomáš, Karafiát, Martin, Burget, Lukáš, Černocký, Jan H., and Khudanpur, Sanjeev (2010). Recurrent neural network based language model. In *Proceedings of the 11th Annual Conference of the International Speech Communication Association*, pages 1045–1048.

Miller, George A. (1957). Some effects of intermittent silence. *The American Journal of Psychology*, **70**(2), 311–314.

Mitzenmacher, Michael (2003). A brief history of generative models for power law and lognormal distributions. *Internet Mathematics*, **1**(2), 226–251.

Montemurro, Marcelo A. (2001). Beyond the Zipf-Mandelbrot law in quantitative linguistics. *Physica A: Statistical Mechanics and its Applications*, **300**, 567–678.

Montemurro, Marcelo A. and Pury, Pedro A. (2002). Long-range fractal correlations in literary corpora. *Fractals*, **10**(4), 451–461.

Montemurro, Marcelo A. and Zanette, Damian (2002). New perspectives on Zipf's law: from single texts to large corpora. *Glottometrics*, **4**, 86–98.

Moradi, Hamid, Grzymala-Busse, Jerzy W., and Roberts, James A. (1998). Entropy of English text: Experiments with humans and a machine learning system based on rough sets. *Information Sciences*, **104**, 31–47.

MRC Psycholinguistic Database (1987). http://websites.psychology.uwa.edu.au/school/ MRCDatabase/uwa_mrc.htm, accessed in October 2020.

Nabeshima, Terutaka and Gunji, Yukio-Pegio (2004). Zipf's law in phonograms and Weibull distribution in ideograms: comparison of English with Japanese. *Biosystems*, **73**(2), 131–139.

Nobesawa, Shiho, Tsutsumi, Junya, Jiang, Sun D., Sano, Tomohisa, Sato, Kengo, and Nakanishi, Masakazu (1996). Segmenting sentences into linky strings using d-bigram statistics. *The 16th International Conference on Computational linguistics*, pages 586–591.

Oshima-Takane, Yuriko and MacWhinney, Brian (1995). *CHILDES manual for Japanese*. Montreal : McGill University, Nagoya: Chukyo University.

Papineni, Kishore, Roukos, Salim, Ward, Todd, and Zhu, Wei-Jing (2002). BLEU : a method for automatic evaluation of machine translation. *Proceedings of the 40th Annual Meeting of the Association for Computational Linguistics*, pages 311–318, Pennsylvania.

Pascanu, Razvan, Tomas, Mikolov, and Yoshua, Bengio (2013). On the difficulty of training recurrent neural networks. *Proceedings of the 30th International Conference on Machine Learning*, volume 28, pages 1310–1318.

Peng, Chung-Kang, Buldyrev, Sergey V., Havlin, Shlomo, Simons, Michael, Stanley, H. Eugene, and Goldberger, Ary L. (1994). Mosaic organization of DNA nucleotides. *Physical Review E*, 49(2), 1685–1689.

Petruszewycz, Micheline (1973). L'histoire de la loi d'Estoup-Zipf: documents. *Mathématiques et Sciences Humaines*, 44, 41–56.

Piantadosi, Steven T. (2014). Zipf's word frequency law in natural language: A critical review and future directions. *Psychonomic Bulletin & Review*, 21(5), 1112–1130.

Piantadosi, Stegen T., Tily, Harry, and Gibson, Edward (2011). Word lengths are optimized for efficient communication. *Proceedings of the National Academy of Sciences*, 108(9), 3526–3529.

Pipiras, Vladas and Taqqu, Murad S. (2017). *Long-Range Dependence and Self-Similarity*. Cambridge University Press.

Pitman, Jim (2006). *Combinatorial Stochastic Processes*. Springer.

Plunkett, Kim and Strömqvist, Sven (1992). The acquisition of Scandinavian languages. In D. I. Slobin, editor, *The Crosslinguistic Study of Language Acquisition*, volume 3, pages 457–556. Lawrence Erlbaum Associates, Inc.

Pruessner, Gunnar (2012). *Self-Organized Criticality : Theory, Models, and Characterisation*. Cambridge University Press.

Rebino, Carl (2021). Reduplication. https://wals.info/chapter/27, accessed in 2021.

Reif, Frederick (1965). *Fundamentals of Statistical and Thermal Physics*. McGraw-Hill.

Ren, Geng, Takahashi, Shuntaro, and Tanaka-Ishii, Kumiko (2019). Entropy rate estimation for English via a large cognitive experiment using Mechanical Turk. *Entropy*, 21(12):1201.

Rényi, Alfréd (1961). On measures of entropy and information. *Proceedings of the Fourth Berkeley Symposium on Mathematics, Statistics and Probability*, pages 547–561.

Robinson, Frank, N. H. (1974). *Noise and Fluctuations in Electronic Devices and Circuits*. Oxford University Press.

Rondal, Jean A. (1985). *Adult-child interaction and the process of language acquisition*. Praeger Publishers.

Ryabko, Boris (2010). Applications of universal source coding to statistical analysis of time series. In I. Woungang, S. Misra, and S. C. Misra, editors, *Selected Topics in Information and Coding Theory*, Series on Coding and Cryptology, pages 289–338. World Scientific Publishing.

Saffran, Jenny R. (2001). Words in a sea of sounds : the output of infant statistical learning. *Cognition*, 81, 149–169.

Santhanam, M. S. and Kantz, Holger (2005). Long-range correlations and rare events in boundary layer wind fields. *Physica A*, 345, 713–721.

Schümann, Thomas and Grassberger, Peter (1996). Entropy estimation of symbol sequences. *Chaos*, 6(3), 414–427.

Segui, Juan, Mehler, Jacques, Frauenfelder, Uli, and Morton, John (1982). The word frequency effect and lexical access. *Neuropsychologia*, 20, 615–627.

Serrano, M.Ángeles, Flammini, Alessandro, and Menczer, Filippo (2009). Modeling statistical properties of written text. *PLoS One*, 4(4):e5372. e:5372.

Shannon, Claude E. (1948). A mathematical theory of communication. *The Bell System Technical Journal*, 27, 379–423, 623–656.

Shannon, Claude E. (1951). Prediction and entropy of printed English. *The Bell System Technical Journal*, 30, 50–64.

Shannon, Claude E. (1959). Coding theorems for a discrete source with a fidelity criterion. *IRE National Convention Record*, 4, 142–163.

Shumway, Robert H. and Stoffer, David S. (2011). *Time Series Analysis and Its Applications: With R Examples (3rd edition)*. Springer.

Simon, Herbert A. (1955). On a class of skew distribution functions. *Biometrika*, **42**(3–4), 425–440.

Smith, H. Fairfield (1938). An empirical law describing heterogeneity in the yields of agricultural crops. *The Journal of Agricultural Science*, **28**, 1–23.

Smoczyńska, Magdalena (1985). The acquisition of Polish. In D. I. Slobin, editor, *The Crosslinguistic Study of Language Acquisition*, chapter 6, pages 595–686. Lawrence Erlbaum Associates, Inc.

Steinert-Threlkeld, Shane and Szymanik, Jakub (2019). Learnability and semantic universals. *Semantics and Pragmatics*, **12**, Article 4.

Stolcke, Andreas (2002). SRILM- an extensible language modeling toolkit. *Proceedings of The 7th International Conference on Spoken Language Processing*, pages 901–904.

Stumpf, Michael P. H. and Porter, Mason A. (2012). Critical truths about power laws. *Science*, **335**, 665–666.

Swadesh, Morris (1971). *The origin and Diversification of Language*. Aldine-Atherton.

Swadesh list (2021). Swadesh list. https://en.wikipedia.org/wiki/Swadesh_list, accessed in 2021.

Takahashi, Shuntaro and Tanaka-Ishii, Kumiko (2017). Do neural nets learn statistical laws behind natural langauge? *PLoS One*, **12**(12):e0189326.

Takahashi, Shuntaro and Tanaka-Ishii, Kumiko (2019). Evaluating computational language models with scaling properties of natural language. *Computational Lingusitics*, **45**, 481–513.

Takahira, Ryosuke, Tanaka-Ishii, Kumiko, and Dębowski, Łukasz (2016). Entropy rate estimates for natural language—a new extrapolation of compressed large-scale corpora. *Entropy*, **18**(10):364.

Tanaka-Ishii, Kumiko (2018). Long-range correlation underlying childhood language and generative models. *Frontiers in Psychology*. Section Quantitative Psychology and Measurement, **9**:01725.

Tanaka-Ishii, Kumiko and Aihara, Shunsuke (2015). Computational constancy measures of texts: Yule's K and Rényi's entropy. *Computational Linguistics*, **41**, 481–502.

Tanaka-Ishii, Kumiko and Bunde, Armin (2016). Long-range memory in literary texts: On the universal clustering of the rare words. *PLoS One*, **11**(11), e0164658.

Tanaka-Ishii, Kumiko and Ishii, Yuichiro (2007). Multilingual phrase-based concordance generation in real-time. *Information Retrieval*, **10**, 275–295.

Tanaka-Ishii, Kumiko and Jin, Zhihui (2008). From phoneme to morpheme: Another verification in English and Chinese using corpora-. *Studia Linguistica*, **62**(2), 224–248.

Tanaka-Ishii, Kumiko and Kobahashi, Tatsuru (2019). Addendum: Another explanation about the bounds of the Taylor exponent. *Journal of Physics Communications*, **3**(8):089401.

Tanaka-Ishii, Kumiko and Kobayashi, Tatsuru (2018). Taylor's law for linguistic sequences and random walk models. *Journal of Physics Communications*, **2**(11):115024.

Tanaka-Ishii, Kumiko and Takahashi, Shuntaro (2021). A comparison of two fluctuation analyses for natural language clustering phenomena: Taylor vs. Ebeling & Neiman methods. *Fractals*, **29**(2). in press.

Tanaka-Ishii, Kumiko and Terada, Hiroshi (2011). Word familiarity and frequency. *Studia Linguistica*, **65**(1), 96–116.

Taylor, Robin, A.J. (2019). *Taylor's power law : order and pattern in nature*. Academic Press.

Taylor, Lionel R. (1961). Aggregation, variance and the mean. *Nature*, **189**(4766), 732–735.

Teh, Yee Whye (2006). A hierarchical Bayesian language model based on Pitman-Yor processes. In *Proceedings of the 21st International Conference on Computational Linguistics and 44th Annual Meeting of the ACL*, pages 985–992.

The Speech Group at CMU (1998). The CMU pronouncing dictionary version 0.6. http://www.speech.cs.cmu.edu/cgi-bin/cmudict.

Thom, René (1974). *Modèles mathématiques de la morphogenèse: recueil de textes sur la theorie des catastrophes et ses applications*. Paris Union générale d'éditions. "Mathematical Models of Morphogenesis" by Brookes, W.M. and Rand, D. published from Ellis Horwood limited.

Thurner, Stefan, Hanel, Rudolf and Klimek, Peter. (2018) *Introduction to the Theory of Complex Systems*. Oxford University Press.

Tian, Ran (2020). Semantic space of language. *Mathematics Seminar*, **701**, 25–29. in Japanese.

Tian, Ran, Okazaki, Naoyuki, and Inui, Kentaro (2017). The mechanism of additive composition. *Machine Learning*, **106**(7), 1083–1130.

Tomasello, Michael (1999). *The Cultural Origins of Human Cognition*. Harvard University Press.

Tomasello, Michael (2003). *Constructing a Language: A Usage-Based Theory of Language Acquisition*. Harvard University Press.

Torre, Iván González, Luque, Bartolo, Lacasa, Lucas, Luque, Jordi, and Fernández-Fernández, Antoni (2017). Emergence of linguistic laws in human voice. *Scientific Reports*, **7**, 43862.

Trefán, Györy, Floriani, Elena, West, Bruce J., and Grigolini, Paolo (1994). Dynamical approach to anomalous diffusion: Response of Lévy processes to a perturbation. *Physical Review E*, **50**, 2564–2579.

Tsallis, Constantino (1988). Possible generalization of Boltzmann-Gibbs statistics. *Journal of Statistical Physics*, **52**(1–2), 479–487.

Turcotte, Donald L. (1997). *Fractals and Chaos in Geology and Geophysics*. Cambridge University Press.

van der Hulst, Harry (2008). On the question of linguistic universals. *The Linguistic Review*, **25**, 1–34.

van Egmond, Mariolein (2018). On the topic of Zipf's law in people with schizophrenic disorders. ph.D. thesis.

van Leijenhorst, D. C. and van der Weide, Theo P. (2005). A formal derivation of Heaps' law. *Information Science*, **170**, 263–272.

von Fintel, Kai and Matthewson, Lisa (2008). Universals in semantics. *The Linguistic Review*, **25**, 139–201.

von Humboldt, Wilhelm (1836). *On language: On the diversity of human language construction and its influence on the mental development of the human species*. Cambridge University Press. English translation by Peter Heath, edited by Michael Losonsky, 2nd edition, 1999.

Voss, Richard F. (1992). Evolution of long-range fractal correlations and 1/f noise in DNA base sequences. *Physical Review Letters*, **68**(25), 3805–3808.

Willinger, Walter, Alderson, David, Doyle, John C., and Li, Lun (2004). More normal than normal : Scaling distributions and complex systems. *Proceedings of the 2004 Winter Simulation Conference*, pages 130–141. IEEE.

Yamasaki, Kazuko, Muchnik, Lev, Havlin, Shlomo, Bunde, Armin, and Stanley, H.Eugene (2005). Scaling and memory in volatility return intervals in financial markets. *Proceedings of the National Acaddemy of Sciences*, **102**(26), 9424–9428.

Yang, Zhilin, Dai, Zihang, Salakhutdinov, Ruslan, and Cohen, William W. (2018). Breaking the softmax bottleneck : A high-rank RNN language model. *Proceedings of International Conference on Learning Representations*, Vancouver.

Yule, George Udny (1944). *The Statistical Study of Literary Vocabulary*. Cambridge University Press.

Zanette, Damián H. and Montemurro, Marcelo A. (2005). Dynamics of text generation with realistic Zipf's distriubtion. *Journal of Quantitative Linguistics*, **12**(1), 29–40.

Zipf, George K. (1945). The meaning-frequency relationship of words. *The Journal of General Psychology*, **33**, 251–256.

Zipf, George K. (1949). *Human Behavior and the Principle of Least Effort : An Introduction to Human Ecology*. Addison-Wesley Press.

Index

© The Author(s) 2021
K. Tanaka-Ishii, *Statistical Universals of Language*, Mathematics in Mind,
https://doi.org/10.1007/978-3-030-59377-3

Printed in the United States
by Baker & Taylor Publisher Services